The Biometric Border World

Since the 1990s, biometric border control has attained key importance throughout Europe. Employing digital images of, for example, fingerprints, DNA, bones, faces or irises, biometric technologies use bodies to identify, categorize and regulate individuals' cross-border movements.

Based on innovative collaborative fieldwork, this book examines how biometrics are developed, put to use and negotiated in key European border sites. It analyses the disparate ways in which the technologies are applied, perceived and experienced by border control agents and others managing the cross-border flow of people, by scientists and developers engaged in making the technologies, and by migrants and non-governmental organizations attempting to manoeuvre in the complicated and often-unpredictable systems of technological control.

Biometric technologies are promoted by national and supranational authorities and industry as scientifically exact and neutral methods of identification and verification, and as an infallible solution to security threats. The ethnographic case studies in this volume demonstrate, however, that the technologies are, in fact, characterized by considerable ambiguity and uncertainty and subject to substantial subjective interpretation, translation and brokering with different implications for migrants, border guards, researchers and other actors engaged in the border world.

Karen Fog Olwig is Professor in the Department of Anthropology at the University of Copenhagen, Denmark.

Kristina Grünenberg is Associate Professor in the Department of Anthropology at the University of Copenhagen, Denmark.

Perle Møhl is Researcher at CAMES – Copenhagen Academy for Medical Education and Simulation, Denmark.

Anja Simonsen is Assistant Professor in the Department of Anthropology at the University of Copenhagen, Denmark.

Routledge Studies in Anthropology

For more information about this series, please visit: www.routledge.com/
Routledge-Studies-in-Anthropology/book-series/SE0724

The Biometric Border World

Technologies, Bodies and Identities
on the Move

**Karen Fog Olwig, Kristina Grünenberg,
Perle Møhl and Anja Simonsen**

Routledge
Taylor & Francis Group

LONDON AND NEW YORK

First published 2020
by Routledge
2 Park Square, Milton Park, Abingdon, Oxon OX14 4RN

and by Routledge
52 Vanderbilt Avenue, New York, NY 10017

Routledge is an imprint of the Taylor & Francis Group, an informa business

First issued in paperback 2021

British Library Cataloguing-in-Publication Data
A catalogue record for this book is available from the British Library

Library of Congress Cataloging-in-Publication Data
A catalog record for this book has been requested

ISBN: 978-0-367-19958-6 (hbk)
ISBN: 978-1-03-208673-6 (pbk)
ISBN: 978-0-367-80846-4 (ebk)

Typeset in Times New Roman
by Apex CoVantage, LLC

Contents

vi *Contents*

Figures

Preface

This book brings together research on migration, technology, policy, visuality and the body by exploring how these fields of investigation meet and intersect in practice in the highly topical and politically charged border world where biometric technologies have become a central method of controlling migration and mobility.

Based on anthropological fieldwork, the book focuses on the daily activities and experiences of different actors in border crossing and avoids the top-down approach in which human agency is marginalized by state apparatuses and their technologies of control. It thereby shows that human activities and choices are the central driving forces. People may be subject to technological control, but this does not take away their capacity to think, invent, move and defy policies of control. State apparatuses may be presented as well-oiled machines, but in practice, they are complex, wide-ranging assemblages of often conflicting agendas, practices and aspirations. Technologies may seem to be scientific, efficient and objective modes of administration and control, but they are made by humans and do not work in mechanical isolation, without human intervention and interpretation. Our joint endeavour has therefore been governed by a wish to look more deeply into the many daily human and technological processes involved in managing Europe's borders, whether they involve attempts to circumvent them, develop them, protect them or avoid being immobilized by them.

When we began the research project at the Department of Anthropology, University of Copenhagen, we had all worked earlier in related fields – Karen, on migration and kinship studies; Kristina on migration, medical anthropology and the body; Perle on identity, everyday communication and visual technologies; and Anja on migration and uncertainty. The research group in the department titled Migration & Social Mobility, including Karen, Anja and Kristina, had been working with questions of legal and fake documents in the context of migration and border crossing. Kristina had become interested in biometric technologies and 'bodily truths' through her earlier studies on complementary and alternative medicine, while Perle was developing a field of visual and sensory research in relation to collaborative projects, as well as scoping regimes of surveillance. Combining these interests made us think about the various types of social and political practices that are set in motion when it is no longer an external document but the digital image of a finger or a face that determines a migrant's legal rights, status and potential mobility. At the same time, we sensed increasing social anxiety concerning, on the one hand, the use of biometric technologies in surveillance and, on the other, a spreading fear of a 'threat from the South' from mass migration. This fear fuelled the already-existing effort to build tighter borders by applying biometric technologies to borders and migration control. Motivated by these dynamics, we set out to formulate a collaborative research proposal, *Biometric Border Worlds*, joining forces to investigate the intersection between migratory movements and biometric technologies. This book is the joint result.

Names, places, translations and illustrations

With a few exceptions, all names of persons and places used in the book are pseudonyms. All interviews and conversations carried out in German, French, Spanish and Danish, and all Danish texts and documents have been translated by us. Interviews in Somali and Tigrinya were conducted with the help of interpreters. Except when otherwise mentioned, all photographs and illustrations are of our own making.

Acknowledgements

All academic writing is an outcome of collaboration in some shape or form. Indeed, it can be difficult to disentangle one's own insights from those of others. Apart from the collaborative endeavour of the four of us, this book is also an outcome of exchanges with our various interlocutors, who generously shared their time, ideas and practices with us, as well as with colleagues who provided valuable inputs to presentations and drafts during the project.

Kristina Grünenberg: I would like to thank the researchers and research managers at the ID Technology Lab and the Biometric Identification Lab for sharing their daily work, their fun conversations and their patience in putting across the intricacies of biometrics to a social scientist; researchers from other European biometric laboratories with whom I engaged during the project; Peter, Professor of Biometrics, for his openness and for facilitating initial contacts and sharing interesting and lively conversations; staff members from biometric vendors Accenture, Gemalto and Stargates for interesting inputs from an industry point of view; the members of the CEN standardization committee on biometrics for letting me participate in their meetings; and countless others involved in what has become known as the 'biometric community'.

Perle Møhl: I would like to thank the Copenhagen Police Border Control Department and the police officers and civil border guards who allowed me to participate in their daily routines, sensory work, briefings and teaching sessions, with a special thanks to JS; the Special Border Police Unit, Southern Jutland Police; Gibraltar Borders & Coast Guard Agency, for allowing me into the entrails of the airport and land border control centres; the Guardia Civil in Ceuta for letting me spend time in their central surveillance centre and participate in their patrols; Lotte Høgh, for consultation on entering the world of the Danish police;

Birgitte Refslund Sørensen and Atreyee Sen for assistance in manoeuvring in a world of complex loyalties. Very heartfelt thanks go to Maite Perez, 'Inmigrantes San Antonio Ceuta', and to the many migrants who had made it across the Ceuta border fence and explained to me their technological skills, pasts and dreams for the future.

Anja Simonsen: I would like to thank the women and men *en route*, Somalis and others, new and old friends, for including me in your often tough and painful journey. Thank you for your warmth, hospitality and kindness: *Ragga iyo dumarka soomaliyeed iyo kuwa aan ahaynba ee ku jirey safarka tahriibka, saaxiibadaydii hore iyo kuwa cusuba waxaan aad iyo aad idiinkaga mahad celinayaa sida aad iiga qayb geliseen safarkii adkaa ee isla markaana xanuunka badnaa. Waxaan aad idiinkaga mahad celinayaa diiranaanta iyo debecsanaanta ka muuqatey sida aad ii so dhowayseen.* Others I would like to thank include the women and men who are part of the Italian social welfare system for undocumented migrants, asylum-seekers and refugees, involving social workers, forensic staff, psychiatrists, non-governmental organization (NGO) volunteers, local Italians involved in charity work, lawyers and staff at the SPRAR centres and reception centres located all over Italy. Thank you for introducing me to your line of work, for sharing your extensive knowledge and for your hospitality. I especially thank Assistant Professor Luca Ciabarri for engaging in a continuous dialogue and sharing his extensive knowledge and sharp analytical insights with me.

Karen Fog Olwig: I would like to thank staff at the following Danish institutions who participated in interviews: the Immigration Service, the Ministry of Foreign Affairs, Folketinget (the parliament), the Department of Forensic Medicine at the University of Copenhagen and a municipal dental care centre for children. I also thank present and former volunteers and staff at formal and informal Danish NGOs as well as staff at Danish embassies abroad, who not only participated in interviews but, in many cases, also invited me to observe their work. Finally, I extend a special thanks to the refugees and their families who made an invaluable contribution to the research by sharing their experiences and knowledge with me, despite their often-difficult circumstances.

Together, we wish to extend our gratitude to Ayo Wahlberg, for inspirational comments and discussions on the project application; the participants in and commentators at the international Biometric Border Worlds workshop, Copenhagen 2017, and at other seminars where we presented papers; Mikkel Rytter, Nigel Rapport, Kenneth Olwig and Vered Amit for their careful reading of different versions of the book manuscript; Heather Horst, Gregory Feldman, Ruben Andersson, Johan Lindquist, Zachary Whyte and Martin Lemberg for comments on earlier papers; and the members of the research groups Migration & Social Mobility and Technology & Political Economy at the Department of Anthropology, University of Copenhagen, for their inputs, fruitful discussions and constructive feedback.

Finally, we would like to extend a warm thank you to the Velux Foundation for the funding that made the project possible and, not least, to Henrik Tronier, who supported the project in all stages of its development.

Introduction

The biometric border world

*Karen Fog Olwig, Kristina Grünenberg, Perle Møhl
and Anja Simonsen*

In a laboratory in Southern Europe, scientists and technical experts are hard at
work developing sophisticated technology that can identify individuals on the
basis of bodily attributes, such as blood veins or heartbeats. In Copenhagen Airport, passengers line up at the automated passport control, where technological
installations will verify their identity by checking their faces against the digital
images stored in the chip inserted in their passports. At 'hotspot' border controls
in Italy, people arriving without the legal documents required to enter the country
are fingerprinted and registered, and the information is sent electronically to the
European database, EURODAC. And at a Danish embassy in East Africa, mouth
swabbing is carried out on children in order to perform a DNA analysis that can
determine their eligibility for unification with their families in Denmark.

Since the 1990s, biometric technologies have become an important element
of border control throughout the world. Through digital analyses of body parts
and bodily substances, biometric technologies are used to identify and categorize individuals in order to monitor and control their mobility. This monograph
explores the epistemological foundations and the practical, social and political
implications of the use of biometric technologies for border making in the context of mobility and migration. Based on collaborative anthropological fieldwork
in different sites of border control, we explore the disparate ways in which the
technologies are applied, perceived and experienced by scientists and developers engaged in creating and merchandising the technologies, by border guards,
police and others managing the cross-border flow of people, and by migrants[1] and
non-governmental organizations (NGOs) attempting to negotiate the complicated
systems of control that they enable. The ethnography therefore contributes new
knowledge of how differing kinds of biometric technologies come into being,
the social and the political significance of their development, how they are used
in actual practice, the assumptions attached to their capacities and use and the
identities, practices and relations they enable and/or disable in varying contexts
of mobility and migration.

Biometric technologies are generally presented and promoted as scientifically
based, exact and neutral methods of identification and verification. However, our
ethnographic case studies demonstrate that in practice they are characterized by
considerable ambiguity and uncertainty. In actual border practices they are subject

to substantial subjective interpretation, translation and brokering, with important consequences for border crossers and the societies involved. Thus, while they have been adopted as part of a general global trend towards the greater securitization of borders and the regulation of population movements, they manifest themselves in distinct ways depending on the cultural and social particularities of different places. We approach these border practices as an assemblage of only partially connected actors that operate in, and co-create, an extensive and ubiquitous realm of border control that is no longer confined to the actual place and moment of physically crossing the boundaries between different states but located in the very bodies of the mobile. We therefore argue that the adoption of biometric technologies has been accompanied by the development of an elaborate, omnipresent and obscure border world outside the national order of things that operates in unpredictable ways. In the following ethnography, we show that this is a border world where the legal and political rights associated with citizenship do not necessarily obtain, where the extraordinary has become routinized and where a range of small decisions, tactics and wishful thinking are the order of the day rather than strategies, firm plans for the future and realistic aspirations.

Our ethnography draws especially on two areas of investigation that have gained prominence in anthropology since the 1990s. We have found an overall framework for our research first in the border studies that recognize the significance of looking at the complex mesh of ideologies, policies, controls, imaginaries, practices and relations within which borders are constructed, given meaning and negotiated by an array of disparate actors. Border studies, as noted by Donnan et al. (2018: 344), have developed within a 'historical period accompanied by dramatic political changes that very much relate to the shifting shapes, functions and meanings of political borders'. The early 1990s, when the 'wall' separating Eastern and Western Europe had fallen and global digital communication was emerging, were characterized by notions of a 'borderless world' where 'fixed territorial borders [would be] a remnant of the past' (Donnan et al. 2018: 344). Indeed, with the introduction of the European Schengen Agreement in 1995, passport control was abolished between the signature countries, and border posts were removed or abandoned. This movement was short-lived, however. While opening its borders within Europe, the EU fortified its external borders in Ceuta and Melilla, North Africa, where walls were built (and reinforced several times) to stem African immigration to Europe. Similar walls have been built in other areas of the world by, for example, Israel and the US, and in recent years, border control has been reinstated by a number of countries within Europe itself (Donnan et al. 2018: 345). At the same time, modern technological border control has become thoroughly biometricized. With its broad spatial and temporal control, which transcends both sides of the narrow line of the border, present-day border practices are quite different from those of the past.

The second major area of research that has informed our ethnography is technology studies examining the ways in which artefacts and devices come into being and are used through multiple sociotechnical practices both within and beyond laboratories. In the context of migration research, anthropologists have tended to emphasize technology's enabling qualities. During the 1990s, Arjun Appadurai, for example, pointed out how the modern electronic media was igniting imaginaries concerning the possibilities for a better life in another part of the world, resulting in a 'technoscape' of interactions, exchanges and mobility on a global scale (Appadurai 1996). The simultaneous development of new electronic technologies of communication and information has similarly led to ethnographic studies of the ongoing, long-distance interpersonal contact that these technologies facilitate and the maintaining of established social relations, as well as the navigating of new terrain that such contacts make possible (e.g. see Madianou and Miller 2012; Baldassar 2016; Zijlstra and van Liempt 2017; Dekker et al. 2018; Gillespie et al. 2018). This monograph focuses on the other side of this technoscape – the technologies of border management that have been instituted by destination countries to control mobility and connectivity. By employing biometric technologies, it is now possible to operate not only at the physical border posts. Biometric registration, identification and control can thus be carried out in the multiplicity of sites where (potential) migrants are bodily present, thereby promising to provide a more comprehensive way to 'protect the secure and developed world from the incursions of the poor and insecure' (Pickering and Weber 2006: 9). By focusing on biometric forms of border control which have fundamentally redefined the conditions of border crossing, this ethnography makes an important contribution to this new area of technological border studies that has only recently become a topic of investigation.

Border studies

Classic fieldwork-based anthropology, with its focus on describing and analysing different cultures and societies, was not particularly concerned with the significance of the borders that separated such entities. In fact, borders seemed to be regarded as 'natural' dividing lines or empty spaces between adjacent socio-cultural entities. However, borders were placed at the centre of anthropological enquiry with the 1969 publication of Fredrik Barth's seminal work *Ethnic Groups and Boundaries*, in which he rejected the 'anthropological reasoning' that 'cultural variation is discontinuous'. Pointing to the pervasiveness of mobility, contact and communication between those who are identified with different socio-cultural groups, he argued that borders do not emerge as the result of a vacuum. Rather, they develop through the construction and maintenance of socio-cultural differences, which takes place in an ongoing process enabled by contact rather than isolation (1969: 9–10).

Barth's work has been influential in research on ethnic groups and identities in societies with large immigrant populations. With his insistence that ethnic groups are not permanent socio-cultural entities based on inherent differences but groupings that are shaped and reshaped through social interaction and negotiation, Barth also highlighted the dynamic nature of borders in social life. From this perspective, borders can be seen to be part of a broader anthropological interest in the role of boundaries in social processes, for example, as reflected in the crossing of borders between different social statuses, as marked in rites of passage or in the 'bordering' of local communities as part of constructions of places of belonging (Donnan and Haller 2000: 8). Common to this work has been an emphasis on the symbolic role of boundaries in socio-cultural differentiation and the associated power relations. There is no natural isomorphism between culture and place (Gupta and Ferguson 1997). Rather, places can be attributed with meaning as sites of particular cultural practices and identities. Such siting, and bordering, of culture will often occur within the context of mobility and interaction (Olwig 1997; Olwig and Sørensen 2002).[2]

Although anthropologists have developed an interest in the role of boundaries in social life, they have generally left the study of international borders to historians, geographers, political scientists and other social scientists. A key issue in the historical study of borders is the role of mapping in the establishment of rigid borderlines between different countries. This process can be linked to the 1648 Peace of Westphalia, which ended a long series of European wars and determined that sovereignty is tied to territory and that states have a monopoly of power within their territory (K.R. Olwig 2013: 258). There was a 'distinct metric element' to this idea, related to the configuration of modern cartography (K.R. Olwig, personal communication 08.03.2019). In this context, international borders have therefore tended to be conceptualized as 'the physical and visible markers of a nation-state's scope', and there has been little interest in their cultural and symbolic significance or the social and economic practices and interactions that might take place among people living in areas adjacent to political borders (Donnan and Haller 2000: 8). During the early 1990s, however, an anthropology of international borderlands emerged, spearheaded by Hastings Donnan and Thomas M. Wilson (1999, 2010), who argued that the 'roles of borderlands and borderlanders cannot be inferred from the major narratives of the nation, or from the grand constructions of the state' that often view borders as simple demarcations of where one country stops and another begins. They suggested that borders should be seen as 'sites and symbols of power' that may attain varying significance in time and space. The role of borders 'in national and state histories and contemporary actions' is therefore 'a matter for empirical research' (2010: 14). Concretely they proposed that borders and borderlands be approached as 'countless points of interaction, or myriad places of divergence and convergence at and around the borderline', or, at a more general level, as 'territorial and cultural spaces of negotiation, mix and interaction' that involve top-down, as well as bottom-up, interpretation, decision-making and practice by legal and administrative institutions and agencies, as well as local populations (2010: 7–8). This means, essentially, that borders do not exist only in physical proximity to a borderline. Border-related spaces of 'negotiation, mix and interaction' also penetrate the wider society beyond the material borderline, influencing both public debates and social, political, legal and economic systems.

Since Donnan and Wilson introduced international borders as an important topic of anthropological investigation, the study of borders has developed rapidly. While Donnan and Wilson originally focused mainly on the borderlands surrounding international borderlines, later research has extended this to include border practices in connection with transnational migration. Biao Xiang and Johan Lindquist suggest that it is useful to focus on the increasingly elaborate 'migration infrastructure' that they describe as 'the systematically interlinked technologies, institutions and actors that facilitate and condition mobility' (2014: 122). With this concept they attempt to shift the focus from migration behaviour to how migration is mediated by a range of socio-material constellations entailing 'migrants and non-migrants, humans and non-humans' (2014: 124). A somewhat similar concept can be found in Gregory Feldman's (2011a: 380) notion of 'apparatus', a 'device of population control and economic management composed of otherwise disparate elements' that 'mediates relations between disconnected actors, ranging from technicians, to policy officials, to scientific authorities, and, in this case, migrants trying to enter the European Union'. From a more economic point of view, Ninna Nyberg Sørensen and Thomas Gammeltoft-Hansen draw attention to the significance of the 'migration industry' that has emerged with the commercialization of migration, as disparate actors have realized the business opportunities associated with migrants' wish to move and governments' attempts to 'manage migratory flows' (2013: 2). This industry includes a wide array of actors, ranging from 'small migrant entrepreneurs facilitating the transportation of people, to multinational companies carrying out deportations; and from individual migrants helping others make the journey, to organized criminal networks profiting from human smuggling and trafficking' (2013: 2).

Whereas the previously mentioned studies focus on different dimensions of the border, Ruben Andersson proposes that a more encompassing approach can be developed by viewing the border as an ecology. The notion of ecology, he argues, 'accounts for the feedbacks, negative and positive, created in an unstable system, as well as for the shifting relations within this system' (Andersson 2016: 36). It is therefore possible to view all actors in the border ecology as equally important, potentially fragile and interdependent, whether humans, machines or the social and political environment. In this way, the ecological perspective offers a wider contextual framework that includes the larger socio-material changes that affect the border world. These studies have therefore pointed to the importance of looking at the wider context of the political, economic, administrative activities that today compose, and shape, international borders.

Understood in terms of infrastructure, apparatuses, industries or ecologies, borders have become ubiquitous and extend into the far corners of society. This is even more the case as migration has increased and a growing number of border crossers have come to be categorized as 'irregular migrants' who are always potentially subject to border control and, possibly, deportation. Moreover, portable biometric suitcases carrying different kinds of biometric sensors and software can, in effect, bring the border anywhere at any time. Borders are also increasingly being externalized, for example, in embassies abroad, where DNA testing is carried out to determine the right to family reunification. Borders have also become a social issue of great significance as reflected in the prominence of migrants and refugees in the popular imagination and political debates, whether they are regarded as threats to security and socio-economic stability or as fellow human beings in need who require special help and assistance. They therefore comprise not only the 'interfaces where the border machinery rubs against specific places, people, and structures' (Andersson 2014: 285) but also the much more diffuse, pervasive sphere of everyday life where borders and bordering have attained practical, political, ideological and symbolic significance today. We suggest that the concept of the 'border world' captures this central, omnipresent significance of borders in contemporary society. Thus, while, during the 1990s, Appadurai pointed to the emergence of diffuse electronic technoscapes of global dimension transcending national borders and enabling continuity and connectivity on a global scale, the concept of a border world points to the concomitant, parallel development of biometric technologies attempting to maintain national systems of differentiation, containment and disconnection. However, this border world is not characterized by a coherent master plan conceived and implemented in an efficient, top-down manner by a well-structured migration regime. Rather, we argue, it consists of 'multiple sites' of only partially overlapping and, at times, contradictory quotidian forms of control and decision-making.

Thinking and approaching the border world

We use the concept of 'assemblage' as a critical tool that makes it possible to avoid conceptualizing the border world as a rational, coherent, intentionally built and well-oiled whole while at the same time enabling us to detect how it is composed of partially connected elements, as well as correlated, coercive and conflicting forces. Indeed, for those who address the notion of assemblage, notably Gilles Deleuze and Félix Guattari, the multiple heterogeneous elements and actors of an assemblage may work together and produce specific effects without being driven by any one underlying organizing force or principle (Collier 2006; Deleuze and Guattari 1987). There is no up–down logic, and the multiple actors and opponents involved all form parts of the unstable assemblage. As Mark Salter writes, '[t]he border, be it state or biopolitical, is precisely constituted through repeated decisions of inclusion and exclusion, entry and exit' (2013: 12). This description of the border finds a resonance in Paul Rabinow's definition of an assemblage as a multitude of small decisions: 'assemblages are conglomerations in which something unfolds that "emerges out of a lot of small decisions; decisions that, for sure, are all conditioned, but not completely predetermined"' (Rabinow 2004 in Rheinberger 2009: 7).

What motivates us to employ the notion of assemblage is this understanding of the importance of small decisions and the absence of an efficient, coherent master plan that we have found reflected in our empirical sites. Through our ethnographic lenses, we are interested in following some of the different factors and forces that in each instance motivate and produce particular decisions and acts in the everyday, whether one is a migrant, a border guard, a developer or someone seeking family unification. Such small decisions are guided by a multitude of very different types of motivations, possibilities, urges and economies – forces that cannot be reduced to single perspectives or phenomena or to one-to-one causal explanations. Legal frameworks, physical obstacles, bodily features, social connections, time perspectives and more interweave and interact and can hardly be isolated, as the connections and mutual influences between them shift constantly. Seen from this perspective, the border world is an assemblage of a multitude of small decisions of shifting intensities, interactions and forms. However, the border-world assemblage does not exist as a thing independently of its registration and representation. Indeed, in our interpretation, the border world is first and foremost an analytical construct that arises through the way in which we conceptualize and approach it, much as does an ethnographic field site.

As anthropologists, we approach the border world through sited fieldwork, as described later. The field of study and the ways we approach it therefore become inseparably linked. But if, as Collier notes, the assemblage is 'an alternative to the categories of local and global, which serve to cast the global as abstraction, and the local in terms of specificity' (2006: 400), our collaborative assemblage approach does not imply that we study our particular sites as instantiations of a bigger, more holistic border-world assemblage. What we describe together are our detailed observations of practices and small decisions in the sites where they take place and where some of their direct effects can be perceived, with insights from each other's sites and the particular agendas that govern them. The sites, in other words, come to 'function as key constitutive arenas' (Andersson 2014: 285) of the border world. In this sense, the sites are in themselves assemblages and serve as perspectives, vantage points from which we may look in more detail at connections, processes and small decisions that become imperceptible when we look at them from above or afar.

From this close-up position we also get a feel for the disconnections at work locally, for example, when commercial and public interests collide or when local police forces make decisions that go against European directives. At the same time, however, we discern attempts to present the border world and its local constituents as coherent and effectual – in other words, we recognise that there is a *wish* for and an attempt to operationalize some kind of coherent master plan, despite the many incoherencies and disconnections we perceive. From our perspective, such attempts, whether expressed in, for example, commission work, policy papers, conferences, directives, technical infrastructure or joint operations often take the form of wishful thinking and dreams about the workings and perfections of a united border world that is enforced by sublime technological solutions.

Biometrics in the border world

Contemporary biometric technologies refer to systems of automated recognition and identification of individuals based on their biological and behavioural characteristics. In this book we deal mainly with biometric technologies that draw on biological features, since these are the technologies that are generally employed in practices of bordering and mobility control.

Since the early 2000s, a substantial corpus of scholarly literature on biometric technologies has emerged within the social sciences. Generally, it has critically examined the normative and unquestioned types of classification, categorization and even discrimination on which these technologies are built and the consequences of this for those who are subjected to the technologies (Aas 2006; Gates 2011; Magnet 2011; Wilson 2006; van der Ploeg 1999; Amoore 2009, 2013). Much of this literature has been linked to larger discussions about the concept of security after 9/11 (Amoore 2009, 2013; Dodge and Kitchin 2004; Magnet 2011; Maguire 2009; Maguire et al. 2014). Some of these studies, furthermore, have looked at the historical processes and technological predecessors of biometric inventions, arguing that these technologies cannot be seen in isolation and therefore are not merely a result of a technological evolution 'improving older identification technologies' (Magnet 2011: 153; see also Breckenridge 2014; Maguire 2009). In general, research has raised questions concerning, on the one hand, how biometric technologies have established a kind of 'identity tag', a coded body in which 'biometrics are turning the human body into the universal ID card of the future' (van der Ploeg 1999: 301) and, on the other hand, albeit to a lesser degree, the various ways in which migrants and refugees perceive and deal with these technologies (Kelly 2006; Navaro-Yashin 2007; Kim 2011).

We approach the biometric border world broadly as an assemblage of technologically mediated practices involving migration officials, migrants and asylum-seekers, developers, policy-makers, surveillance video operators, NGOs and now also ourselves as anthropologists. Borders materialize when biometric technologies, such as facial scans and fingerprints, are matched with existing digitally coded body images and fingerprints, as in the biometric gates at the Copenhagen airport. They are generally used to authorize access to certain places and services for certain people and to deny this access to others. Different border regimes, furthermore, are invoked when biometric technologies, such as fingerprinting, in principle fix refugees in time and space, or DNA analysis determines individuals' family relations, and thus their legal right to family reunification as migrants and refugees, even though kinship is often fluid, negotiable and attributed with varying meanings in different socio-cultural contexts. Indeed, a central characteristic of biometric technologies, we argue, is their ability to offer seemingly clear identifications and registrations of border crossers and thus give the appearance of enabling efficient management of a complex border world in the face of what is perceived as massive international migration. Biometric technologies, in other words, can be seen to help governments 'reinforce the symbolic and discursive importance of borders as impenetrable barriers' at a time when borders are 'being reinterpreted, fragmented and manipulated' (Pickering and Weber 2006: 10). The border, then, is as much biopolitical as it is geopolitical.

The measurable body and the advent of biometric technologies

The story of biometrics is configured by several parallel and uncoordinated narratives of invention and forgetting rather than as a coherent linear tale. The use of body prints and attributes – hands, palms, fingers and feet – as a way of linking objects and documents to particular persons has seemingly existed since ancient cave paintings. It is also known to have taken the form of inky foot- and handprints on sixteenth-century Chinese divorce certificates and other legal documents. While these very early body prints are sometimes mentioned as predecessors of biometrics in publications on biometric technologies, there are disagreements among researchers as to whether they were used for aesthetic and ritual purposes rather than as proxies for particular identities, pointing to some of the controversies in this field (Sommer 2015; The Day Book 1912; Maguire 2009: 10). The move from such direct imprints of bodily features towards biometrics – that is, measuring the biological – implies a series of concomitant conceptual moves, notably thinking in terms of whole bodies that have individual characteristics, individual bodies that are measurable as well as, fundamentally, the notions of quantification, measurement and scale. Such forms of knowledge are related to the emergence of what in Foucauldian terms we could call a political history of bodies, coupled to ideas about territory, measurability and representation.

The invention of biometrics and its systematization have also been associated with physical anthropology. Franz Boas, for example, known as one of the founding fathers of American anthropology, was involved in discussions regarding anthropometry, or the measurement and metrification of the body, in the late nineteenth century (Boas 1893). As an applied form of science, the early work on biometry is generally attributed variously to the French policeman Alphonse Bertillon, who elaborated a system to measure and categorize different parts of the body for criminal records; the Italian professor of psychiatry Cesare Lombroso, who from studies of skull and face structure developed his theory on biologically founded criminal behaviour; the colonial administrator William Herschel, who experimented with the use of fingerprints as signature in colonial India; the Scottish missionary and physician Henry Faulds, who came across fingerprints on ancient pottery in Japan and proposed their use for criminal identification; and the English mathematician and psychologist Francis Galton, who contributed to the elaboration of a general identification system on the basis of fingerprints (Finn 2005, 2009; Maguire 2009; Breckenridge 2014).

These figures were all contemporaries in the late nineteenth century who set out, each in their own contexts, to explore how the body and its particular metrics or measurements could serve, for example, to uniquely identify people and possibly reveal their likely intentions and inclinations. Together with a host of other researchers, assistants, criminologists, administrators and funding agencies, they developed technologies of human – particularly criminal – identification and classification on the basis of the body. Their efforts were motivated by a fundamental interest in establishing unambiguous ties between particular bodies and identities and were at the time mostly coupled with interests of law enforcement and the state. In this process, according to Mark Maguire (2009), they 'foresaw and spoke of the possibilities for [contemporary] systems of biometric security' (Maguire 2009: 10; see also Finn 2005, 2009).

Controversies at the time about who actually invented fingerprinting – William Herschel, Henry Faulds or Francis Galton – still persist today and point to the impossibility of providing a single-stringed historical account of biometrics. However, it was Francis Galton who took the credit for the invention within the scientific community with his 1892 book *Finger Prints* (cf. Finn 2009;

Breckenridge 2014). At the time, fingerprint archives were tied to specific physical locations, and a significant amount of manpower was required to produce entries, file them and search through them (Finn 2009). As Jonathan Finn 2009 argues, '[a]lthough in theory police could photograph a seemingly infinite number of fingerprints, in practice this would render the archive unusable. As the archive grew, searches became increasingly difficult and laborious, and positive matches became increasingly improbable' (Finn 2009: 106). Today the capturing and comparison of body prints (hand, palm, finger, etc.), facial features or vein patterns has been automatized, and digital templates are stored in databases that take up little space and can be searched algorithmically by computer processing.

Specific biometric technologies

In the following, we describe some of the most prevalent biometric technologies currently used to establish borders in the context of migration.

Fingerprints

The uniqueness of a fingerprint is established from the patterns on the finger surface known as ridges, valleys or furrows, systematized by Galton and others. These features are still widely analysed in biometric technologies for security and mobility-regulating purposes. In automated biometric border control, the fingerprints are usually 'live scanned' into digital images through the sensing of the finger surface with an electronic scanner. The most conventional approach is for the automated system to capture and compare particular points in these fingertip landscapes, known as 'minutiae points', by transforming them into digital templates that can be stored and compared either with a live version of the fingerprint or with other fingerprints. This comparison subsequently enables or disables, for example, an individual's passage through a border control.

Minutiae point comparison requires high-quality images to produce what are defined as sufficiently accurate results, although this quality is not always obtainable in practice (cf. Maltoni et al. 2009). Some individuals

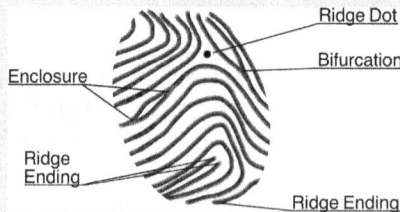

Figure 0.1 Part of fingerprint with minutiae points

Source: © author.

cannot be registered by fingerprint scanners as a consequence of problems such as skin disease, erosion of the ridges due to physical labour, intentional erasure of the prints, or various dry skin- or age-related problems (cf. Magnet 2011). Furthermore, demonstrations of the possibility of circumventing biometric fingerprint sensors with rudimentary materials such as gummy bears and silicone (cf. Matsumoto et al. 2002; Leyden 2002) have led researchers to invent new applications that may serve to protect biometric installations against what is informally known as 'spoofing' ('presentation attacks' in the standardized biometric vocabulary; see Grünenberg forthcoming). Examples of such technological additions include applications that are able to 'detect liveness': pulse, perspiration and skin pores, as well as the extra fingerprints that exist *under the skin* and can thus be used to detect real from fake fingers.

DNA

The notion of a fixed bodily essence that is unique to the individual and that signifies hereditary physical bodily traits and diseases goes back to antiquity (Yapijakis 2017). However, the scientific models substantiating and explaining these ideas were only established in the early 1900s, while the molecule corroborating and sustaining these models, the DNA double helix, was visualized by X-ray diffraction images as late as in the 1950s by the English chemist Rosalind Franklin and her colleagues (Maddox 2003). The systematic use of DNA in forensic science, in screening for hereditary illnesses and in determining biological kinship is even more recent, going back only to the mid-1980s.

As a biometric technology, in migration control, DNA is mainly used to establish biological relations between two (or more) individuals through the comparison of two sets of DNA in order to assess applications for family reunification. The available DNA is sequenced to determine the exact order of the molecules in the DNA strand, and the two sets of DNA profiles are compared. Larger or smaller amounts of DNA material can be analysed, thus augmenting or lowering the statistical probability of the analysis.

While most other biometric technologies that identify people with bodily characteristics can be directly experienced, the molecular scale of DNA makes it an object of imagination and abstract categorization, thus leaving DNA testing to the unique authority, or even monopoly, of science. This makes DNA analysis largely obscure, sometimes even alienating, for most people, and it takes away much of the agency in processes that supposedly provide foolproof evidence of their lives, origins and relations.

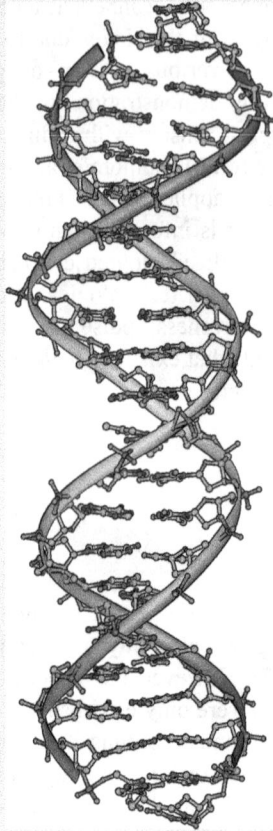

Figure 0.2 DNA double helix

Bone X-rays

X-rays are electromagnetic rays that, when projected onto an object, can visualize its inner structural composition because the rays are absorbed differently by materials with different densities, notably by the calcium in bone structures. The X-ray of a body or body part, coupled with knowledge about normal skeletal development, can be used to assess the age of the person in question, often complemented by X-rays of teeth and a bodily examination.

Forensic pathologists use bone x-rays to make a rough estimate of the age of young asylum-seekers and the children of refugees, generally to determine whether they are legally minors or adults and thus determine their status in the refugee system as well as in questions of family reunification.

As images, X-rays have the authority of photographic images as seemingly direct, unmediated traces of a presence – here, of a structure that cannot be seen with the naked eye. Despite such supposed objectivity, this authority, and its reliability, is coupled with the individual subjective interpretation of the X-ray image, as well as what the forensic specialists themselves acknowledge as 'observer variation' (Shackleton et al. 2004). When it comes to age assessment, for example, in archaeology and forensics, the statistical variations are moreover rather large: the younger the person, the more reliable the assessed age. The approximate ages given in cases of age assessment usually operate with a statistical deviation of several years (Benson and Williams 2008). They are further complicated in cases where the person has been subject to malnutrition and illness, as is often the case when assessing the age of young refugees (Cunha et al. 2009; N. Lynnerup, personal communication 16.11.2016). Furthermore, there is little research on the variations between ethnic groups (Cunha et al. 2009; see also Netz forthcoming).

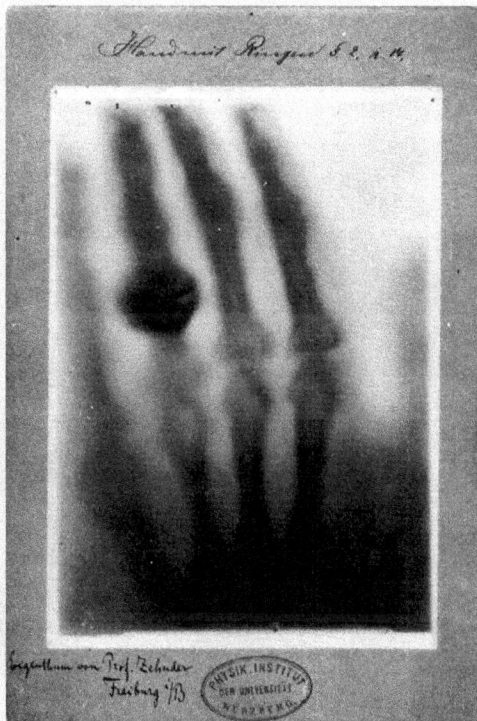

Figure 0.3 The first X-ray (1895) by Wilhelm Röntgen. The image featured the left hand of his wife. Upon seeing the X-ray, Mrs. Röntgen exclaimed, 'I have seen my death.'

Facial recognition

The particularity of the face for use in biometric facial recognition is established by detecting and abstracting faces from image backgrounds with the help of detection algorithms and what are known as facial 'landmarks' (e.g. ears, eyes, eyebrows, nose, facial shape). Subsequently, features defined as salient to a particular face are extracted and stored as a template for later comparison. Described as 'natural and non-intrusive', the most important advantage of using the face for identification and/or recognition, according to scientists in the field, is the fact that, unlike other biological features, faces 'can be captured at a distance and in a covert manner' (Li and Jain 2011: 1).

Facial recognition, like fingerprinting, is used for access control and mobility regulation in a wide variety of settings, from smartphones to biometric border gates. In border control the live image captured in the biometric gate is automatically compared to the image in the passport chip by the

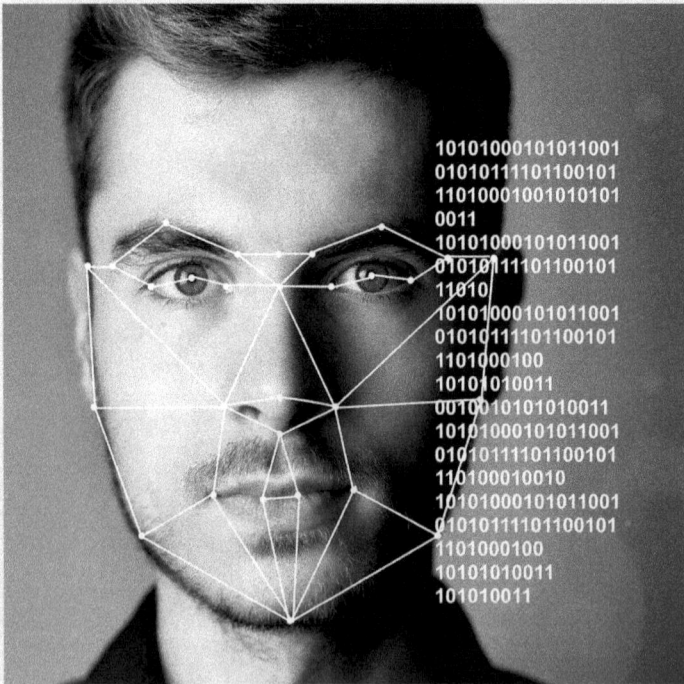

Figure 0.4 Facial recognition template

Source: © author.

facial recognition software. Border guards, in fact, read faces in similar ways, comparing the live face to the ID photo.

Facial recognition is highly dependent on the quality of images, environmental conditions such as light intensity and behavioural issues such as pose, movement, facial expressions and the use of glasses. Furthermore, face detection can be 'spoofed', that is, fooled by make-up, masks and other types of facial reconfiguration.

ID photos

Most biometric technologies, including those described here, either involve direct verification through image analysis (e.g. facial recognition, fingerprints, X-rays, presence detection) or provide information from automated processes, such as chemical analysis, in the form of images. Photographs thus are essential to almost all biometric technologies at some level. They are also themselves defined as a form of biometry (Biometrics Institute 2015) constituting prints of bodies, where bodily and facial features are transposed into assessable, downscaled, often digitized formats through the reflection of light. The ID photo included in most ID documents is no exception. As a biometric technology, it is used, for example, by border guards when they compare an ID photo to the face of the person in front of them and by automated systems of identification, such as facial recognition in Automated Border Control.

The authority of the ID photo is established by both its indexical semiotic qualities, as a seemingly direct unmediated trace of a presence (Møhl 1993; see also Chapter 3), and the scientific and political economy in which its production and use is enrolled.

Future technologies

Biometric technologies are continuously being developed by researchers applying their imagination of possible futures in the context of border crossing, often coupled with ideas of potential security threats. Known modalities are constantly being refined, according to researchers, in order to prevent 'spoofing' and enhance security (Goudelis et al. 2009). Current efforts to improve the technologies of fingerprints and facial recognition include combining them with skin texture recognition and the detection of pulse and patterns under the somatic surface. These body patterns under the skin are captured using, for example, thermal and infrared sensors (see Part I). Palm, finger and wrist veins for purposes of recognition constitute such 'sub-dermal' biometric characteristics that are increasingly used for the purpose of identification in as different contexts as payment for food at the cafeteria of the Copenhagen Business School and boarding domestic airplanes in Korean airports.

Ear recognition, identifying persons from the shape of their outer ear, and heart biometrics, using the particular shape and beat of the heart for identification and verification, are among the newest modalities being developed (Intagliata 2017). Furthermore, researchers have recently tested the possible use of brainwave frequencies as biometric identifiers (Burt 2018). In addition to these technologies there are various 'behavioural biometrics', including voice recognition and gait analysis, that are already quite advanced, and gesture, lip movement and odour are also being explored as sources of biometric identification. As yet, only some of these technologies have been tested outside laboratories.

Migration, mobility and smart borders in a European context

Since the 1990s, biometric technologies have played an increasingly important role in border control. This development has been accelerated by the heightened anxieties and feelings of insecurity and the subsequent demands for enhanced security, since the 9/11 terror attacks and the increased focus on external threats. In this context, biometric technologies were envisaged as important solutions, and biometrics have become a global billion-dollar industry, with an increasing number of actors (e.g. commercial companies, consultancies, researchers, security agencies) turning to the development and sale of biometric technologies (cf. Amoore 2013). Since 2008 the EU has worked on the implementation of biometric border installations across its territory. And in 2009, biometric passports containing fingerprints and facial images became mandatory for most EU citizens. In February 2014, the EU completed the 'smart borders package', which is expected to become fully operational in 2020 (Consuegra 2019). It is introducing biometric technologies in border control in order to 'improve the management of the external borders of the Schengen Member States, fight against irregular immigration and provide information on overstayers, as well as facilitate border crossings for pre-vetted frequent third country national (TCN) travellers' (EU Migration and Home affairs web 2019).

Control of external EU borders has been implemented largely by the European agency for the coordination of joint border operations, Frontex, today also called the European Border and Coast Guard Agency. The agency was established in 2004 with a mission to 'ensure safe and well-functioning external borders providing security' through collaboration between the EU member states (Frontex 2019a). Today, Frontex concentrates mostly on the southern borders of Europe (Collyer et al. 2012: 409) by analysing migratory patterns and cross-border criminal activities, coordinating and organizing joint operations and rapid border interventions, coordinating and providing training for the national border forces and supporting EU member states in their ongoing control of border crossers and their implementation of forced returns of people without a legal right to remain in the EU (Frontex 2019b). While Frontex originated in the European project to ensure free movement within its borders, it has become a central player in managing Europe's external borders and today has a gigantic budget of 320,198,000 euros (Frontex 2018).[3]

The huge investments in European border control must be seen in the light of a growing concern with the influx of migrants and refugees. In 2016, the number of men, women and children seeking asylum in Europe peaked when an estimated 362,000 people crossed the Mediterranean Sea, 181,400 of them arriving in Italy (UNHCR 2019a). This was a relatively small number, given the estimated total of 65.6 million displaced people in the world, largely as a result of civil war (e.g. in Syria and South Sudan) and continuing conflicts and instability (e.g. in Afghanistan and the Horn of Africa; UNHCR 2019b). Nevertheless, in Europe the influx of people was framed as a 'migration crisis', and various migration management tools were introduced to bring this population movement to a halt. Border controls were re-established (temporarily) between several of the member states of the EU and stricter controls instituted at external EU borders.

In addition to the increased border control by agencies such as Frontex, agreements have been made with countries that have served as points of entry into Europe, such as Turkey, Libya and Morocco, with the aim of stopping the flow of migrants.[4] Furthermore, hotspot border control has been implemented in five areas in Italy and Greece, the two European countries receiving by far the largest number of migrants. The hotspot approach was formulated by the European Commission in April 2015 (European Union Agency for Fundamental Rights (FRA) 2018), one of its main objectives being to identify and register all migrants arriving at the EU's borders (European Commission 2015a) through rapid pre-identification, registration, photo and fingerprinting operations (Capitani 2016: 4). Finally, as a growing number of refugees have received asylum, some European states have introduced restrictions on refugees' right to reunification with family left behind by, for example, instituting lengthy waiting periods for the right to apply (Muižnieks 2017).

Biometric technologies are used as a tool not only to control irregular movement but also, to some extent, to provide care for refugees in the form of housing, health and legal services. Fingerprinting has been at the core of border control for many years, and all European passports now contain fingerprints and a photograph that can be used to identify travellers. Indeed, major investments in automated biometric control are being made at many airports and land borders. Fingerprints, and from 2018 facial images, also feature in EURODAC, the biometric database that stores biometric data on those who have entered the EU on 'an irregular basis', that is, without the required documentation (Schuster 2011: 404). Such irregular migrants include asylum-seekers, whose registration in EURODAC falls under the Dublin regulation of 1997, which specifies that the EU member state where an asylum-seeker is first registered is responsible for processing the application for refuge (Hurwitz 1999). This regulation is intended to prevent individuals from seeking asylum in several European countries, but since the vast majority of recent asylum-seekers have entered Europe in Italy or Greece, it has placed a huge responsibility on these two countries with regard to the reception and administration of asylum cases. DNA analysis and biometric age assessment have also become significant in the control of immigration through family reunification when refugees do not have what are considered trustworthy documents proving their status in the family (Heinemann et al. 2015: 2). As migration management tools, these biometric technologies are thus employed not only to identify travellers and monitor their movement but also to divide people into 'legitimate' travellers (primarily from the EU and the Global North), whose journeys must be as seamless as possible, and 'illegitimate' border crossers (mainly from the Global South), whose onward movement must be stopped.

While fingerprinting, DNA analysis and X-rays are established technologies that have been in use for many years, automated facial recognition and iris scanning are much more recent technologies that, as noted, are used at automated border control gates. These fairly recent technologies have also been implemented in, for example, refugee camps in an attempt both to provide aid and to register arrivals more efficiently (Jacobsen 2017). Furthermore, considerable government and private investments are being made with a view to developing new, more sophisticated biometric technologies that can make border control more efficient, comprehensive and secure, curbing the movement into Europe of so-called irregular migrants.

Sites and methods

In order to capture the complexities of this biometric border world, with its many different actors negotiating cross-border travel from disparate vantage points, often with contradictory agendas, this monograph has adopted a collaborative, strategically sited ethnographic approach based on anthropological fieldwork in four key sites of border activity. It begins in a laboratory of biometrics and at biometric events and conferences where new technologies are developed, tested, presented and discussed, with the aim of closing potential gaps in the biometric system of control. It moves to the actual physical border crossings, where the flow of people is checked by the border police using visual screening, data technologies and their own senses to identify and stop 'irregular' travellers and where migrants develop and apply their own technological savvy. It then proceeds to the travel routes used by migrants who attempt either to stay clear of biometric registration and identification in order to avoid having their travel plans curtailed or to ensure that they are registered in what they regard to be desirable destination countries. Finally, our ethnographic study ends by examining refugees' endeavours to obtain family reunification after periods of separation due to war or conflict and the role played by the biometric assessment of claimed kin ties. Whereas this approach might indicate that we are embarking on a classical linear top-down tale of biometrics from development to experience, the story of biometric technologies in a migration context does not start or end in science labs, at airports, with migrants *en route* or in the processes of family reunification. A book, however, has its own linear logic and must start somewhere.

With our focus on fingerprinting, vein biometrics, DNA analysis, bone analysis and facial recognition, these four sites show how some of the main biometric technologies are used today in border control in concrete places. In this way they illuminate both some of the general features and the social and historical specificities of local practices in the biometric border world. The mobile body becomes a consistent figure throughout the sites. The body, for example, becomes mobile as part of biometric databases shared between biometric researchers in laboratories, and it also changes shape throughout its movement in the lab. On the border, travelling bodies move through biometric border gates, while the bodies of border guards simultaneously move between different sensory registers. Migrant bodies move toward what is considered safer and better lives or between countries and embassies in order to deliver samples of saliva on foam-rubber buds, body excretions that are, in turn, made mobile when sent for analysis in, for example, Copenhagen.

The sites we explore together and individually can best be described as composed of a series of encounters that we ourselves have defined as four interlinked, 'extended field sites' (Andersson 2014; see also Olwig 2007: 22–24) in the biometric border-world assemblage. Here we wish to present these extended field sites briefly, describing more concretely how we assembled them in practice through our particular modes of inquiry and methods.

In the lab

In their laboratories, biometric researchers are constantly refining the types of biometric technologies in use and exploring and experimenting with new ways in which different body parts and topologies might fruitfully be 'enrolled' for biometric use. The technical work on fingerprints, veins, faces, voice, gait, heartbeats and so on is, however, deeply entwined with social practices and does not take place in isolation from the world beyond the laboratory. Instead, laboratory practices are shaped by scientific protocols and by the ways in which laboratories mobilize and are mobilized by external partners, policy-makers, funding agencies and contractors, including the kinds of interests that are attached to biometric technologies as mediators and regulators of mobility in the border world. From the laboratories, these two chapters therefore engage in a variety of settings, including events organized by biometric interest organizations, biometric and security conferences, lectures on biometrics, a biometric vendor fair and a meeting in the EU standardization committee on biometrics.

The ethnographic focus in this site is, on the one hand, how the researchers come to see their work on bodies and biometrics as meaningful both to themselves and to an imagined future and, on the other hand, the multiple performances, sociotechnical practices and relations that continually actualize the border world in the labs and thus ultimately shape biometric technologies.

On the border

The border world is, as noted earlier, an infinite number of small actualizations taking place in various dispersed sites, as well as consisting of fences, gates and national borders. Studying the integrated border work of border guards, migrants and technologies implies approaching borders as sites of daily interaction and intense work – as workplaces – and studying all the parties and agendas that go into their daily production, maintenance and circumvention.

The chapters in this section take us to three rather different EU/Schengen border settings: passport control in Copenhagen Airport, at Gibraltar International Airport and the land border to Spain and the border fence between Spain and Morocco in Ceuta, a Spanish city enclave on the Moroccan coast. All three are Schengen borders, but the practical settings and conditions under which they work are very different, being, respectively, airports within a national territory receiving passengers arriving by air and physical borders between two national territories with a border fence or a sea providing the line of separation.

This section focuses on the people engaged in border work, whether border guards or migrants: how they deploy and develop their intuition, their interpretive skills and their senses, notably their vision, and how they *learn to see*, their use of different kinds of biometric technologies, and the implications of these human-technological interactions for the innumerable small daily decision-making processes taking place around the physical border.

En route

African migration trajectories usually cross the southern borders of Europe and continue towards Northern Europe, as people search for better social and economic opportunities. Often, however, their journeys are cut short by European border control, despite numerous attempts to travel onward. Focusing on Somali women and men who are or have been stranded in Italy, this section examines how EU migration management initiatives, such as the mandatory biometric registration of potential asylum-seekers' fingerprints upon their arrival in Italy, have been implemented and their consequences for the Somalis' continued mobility. The section is based on fieldwork focusing on social and human-technological relations as they are played out in practice at the border sites and in the Somalis' temporary places of refuge: the churches, public housing centres, the parks and the occupied buildings where they eat and sleep among other Somalis engaged in wide-ranging trajectories in search of a better life. These relations include border police officers, Italian social workers, Italian employees and volunteers and other Somalis, whether brokers, strangers, friends or kin living in or outside Italy.

These two chapters examine how Somali women and men experience being biometrically registered through fingerprinting in this particular part of the European border world and the kinds of networks they turn to in their search for a secure life.

In the family

Perhaps the most geographically dispersed and socially intrusive site in the border world concerns the many actors involved in the biometric assessment of family reunification among refugees, through DNA testing and X-rays of bones and teeth. This extended field site comprises a broad range of actors, including refugee families that have become involuntarily separated and are living under disparate conditions in terms of security and access to social and economic resources; NGOs and lawyers who assist, often on a voluntary basis, family members seeking reunification; officials in immigration systems receiving and processing applications for reunification; staff at hospitals and embassies conducting biometric tests and analyses of applicants; politicians instituting shifting policies; and various media reporting, as well as general public debates on family reunification.

Based on research in documents and fieldwork in Denmark and East Africa, this final section examines how DNA testing became a central part of the Danish assessment of family reunification, and the ways in which this has influenced refugees' ability to re-establish their family lives and create a new existence for themselves in Denmark. The chapters illuminate the disjointed, contradictory and multiplex ways in which the biometric border world can penetrate into individuals' personal family relations, one of the most intimate spheres of human life.

Methodological challenges and opportunities

Doing ethnographic fieldwork in these areas of border activity clearly poses a number of challenges, such as gaining access, adjusting methods to the particular possibilities and limits associated with the accessible sites, and coming to terms with the ethical issues raised by doing research in environments that are politically highly charged. Taking sides and the nature and sources of anthropological authority are questions that anthropologists continually grapple with, and epistemological discussions of anthropological positionality and partiality remain relevant and timely topics (cf. Armbruster and Laerke 2008; Bauer 2014; David 2002; Hastrup 2004, 2015; Becker et al. 2005; Mosse 2008). Such interrogations become pivotal when collaborating in a dense, highly unstable and multi-perspectival field such as the biometric border world, which is politically, technologically and emotionally engaging. At times, our collaborative approach clashes with our desires to share our new knowledge. This has produced some fundamental questions about the loyalties we uphold during fieldwork with people who are positioned very differently in the border world but who all play important roles and are involved in practices with which we wish to engage.

Considerations about how, or even to what extent, we can share knowledge and with whom are especially pertinent because information from one actor in the biometric border world can have direct implications for others. Through our ethnographic engagement in the biometric border world, we have, for example, gained knowledge about the tactics used by the border police units that try to prevent people without legal documents from crossing borders; about migrants attempting to challenge such tactics when crossing international borders; about the imperfections and uncertainties of biometric technologies; about the necessity of migrants engaging in illegal border crossing in order to gain access to biometric testing for family unification; and much more. Whether to share details about such acquired knowledge with each other, as well as with the world at large, has been a recurring issue that has posed questions as to whether the sharing of privileged knowledge would be a breach of loyalty to the people with whom we have engaged in the field and who have let us into their daily lives and worlds.

There were certain limits to the kinds of observation and participation we were allowed to undertake in our sites, where issues of security and protecting sensitive, intimate and classified information were important ongoing concerns. Despite these issues and questions, in all the sites we carried out classic anthropological fieldwork based on participant observation and formal, as well as informal, qualitative interviews, supplemented by research in legal, policy and technical documents. The resultant ethnography of the biometric border world examines how biometrics are developed, put to use and negotiated in key European border sites. Thus, it analyses the disparate ways in which the technology is applied, perceived and experienced by border control agents and others managing the cross-border flows of people, by scientists and developers engaged in creating and sometimes merchandising the technologies, and by migrants and NGOs attempting to manoeuvre in the complicated and often unpredictable systems of identification and control enabled by the biometric technologies. In sum, in this book, we invite the reader to enjoy a rare view of the biometric border world that can offer insights into a little-known, but increasingly important, aspect of contemporary life.

Notes

1 Throughout the book, we use the term *migrant* as a general category to describe people who travel across borders, with particular reference to those who do so without the necessary documents. This is to underline how such travellers, whether labour migrants or asylum-seekers, move along the same routes, experience the same dangers and risk their lives to reach Europe, despite the different political framings of their movement. When the specific political category is important for an analytical point, terms such as *asylum-seeker*, *undocumented migrant* and *refugee* are employed.
2 Please note that all references in the text that cite simply 'Olwig' refer to publications by Karen Fog Olwig. Publications by Kenneth Robert Olwig will be cited as 'K.R. Olwig'.
3 This is a huge increase since 2006, when Frontex operated with a budget of 19 million euros (Collyer et al. 2012: 408; Frontex Amending Budget 2006, 2007).
4 The EU-Turkey Refugee Statement, signed in March 2016, thus stipulates that Turkey agrees to take back 'irregular migrants' entering Greece in return for considerable economic compensation from the EU (European Council 2016). The 'Italy-Libya Memorandum of Understanding', signed in February 2017 entails Italy providing financial and technical assistance as well as training to Libya in order to strengthen Libyan border controls with the aim of controlling the migration flows through the country (Palm 2017).

Part I

In the laboratory

Kristina Grünenberg

Introducing the site

The ethnography of the biometric border world, which we have set out to study in this book, takes off in a laboratory where biometric technologies are researched and developed.[1] Such contemporary biometric laboratories differ from most people's imaginaries of science laboratories as places where people in white coats work with test tubes, not to mention 'test animals', in sterile research environments. Biometric researchers do not deal with cells, molecules or blood samples. Instead, they enquire into the aptness of new parts of the body as biometric identifiers. They share and build body-image databases, write algorithms and use programming languages to convert them into code that tells the computer how to process such digitalized body images and so on. In short, they develop applications that may be used in different types of systems of identification and verification, ranging from mobile phones to border gates in airports. By the time biometric algorithms, code and programs developed through diverse social and material practices in the labs have been installed in, for example, biometric border gates, these gates come across as objects, or things in their own right. Consequently, as travellers approach such assemblages of computer software and applications, written instructions, cameras, metal, glass and flashing red or green lights in their (supposedly) finished shape, they generally have little idea as to how they operate and of the multiple types of practices that brought these gates to the border.

By focusing on the work that takes place in and around European biometric labs, this first part of the book 'makes visible' some of the meticulous technical and social practices and dynamics that play a role in configuring biometrics as particular border technologies, including how they are co-configured politically and economically. The purpose is to show how the interests, aspirations, technical expertise, small daily decisions and work practices of biometric researchers, as well as their entanglements with a particular field of relations, co-configure biometric technologies and the border world.

Most of the biometric border sites encountered throughout this book are operationalized through scientifically well-known and well-researched biometric technologies (or modalities) that are already in extensive use, such as fingerprints, DNA and facial recognition. In the contemporary labs where I did fieldwork, the concern with these established technologies was mostly related to the need to improve, modify, refine or test them, as well as to find ways to make them more 'spoof-proof' or resistant to 'presentation attacks'.[2] What the researchers saw as most interesting and challenging, however, was the experimentation with and development of new technologies of the body that could be used in the future. Taking its point of departure in vein biometrics, implemented for the first time as a border technology in a South Korean domestic airport in 2018 this part of the volume therefore examines the ways in which newer technologies come into being and are developed further in biometric labs.

Biometric labs

Biometric laboratories affiliated to universities can be found across Europe, often as part of larger departments or research centres focusing on, for example, machine vision and signal analysis, computer science and information security. Most university labs have few permanent members of staff and rely on external funding in order to maintain a size that makes it possible to undertake diverse complex and practical tasks and to generate frontline research. The labs where I did my fieldwork employed between 9 and 12 researchers, depending on the funding situation, the ongoing projects and the tasks that needed solving. Biometrics is a relatively small academic world, and the researchers from the different EU biometric labs generally know one another directly or indirectly. The labs work with both industry, in the form of biometric companies, and various sections of government from the intelligence services to welfare and health departments and ministries both within and outside the EU. Although the larger biometric companies often have their own research and development departments, they hardly ever develop an entire biometric system on their own. In fact, many biometric companies work mainly as integrators of software and hardware produced by others. Such companies may need the basic research provided by researchers, such as particular algorithms, codes and programs, and/or the prestige and trust invested in university research by funding agencies and sources for larger collaborative projects and tenders (including funding from the EU and national governments). In biometric laboratories, researchers will therefore engage rather intensively in commercial and/or externally funded projects but also ensure that this funding allows them to cultivate experimental basic research, which generally constitutes their core interest and is the prime motivation for their work. At the centre of many of these externally funded projects, however, are issues of security, not least state security and border regulation.

In the following, my main focus is on one particular European laboratory, located in a Southern European capital, which I call 'the ID-technology lab'.[3] I use this lab as a basis for a broader description of biometric labs but, in addition, draw on material from a second European laboratory, 'the Biometric Identification Lab', where I also did fieldwork. This is supplemented by knowledge I have gained from other biometric labs through conversations with lab researchers at different events and by participating in research and border security conferences and associational and network meetings. Through all these encounters, I have engaged in what a professor of biometrics referred to as 'the biometric community'.

Introducing the ID-technology laboratory

> *The ID-technology lab is located at the technical university, a rather large campus with red-brick buildings surrounded by green patches of grass, trees and benches. The buildings are made up of endless similar-looking corridors and stairways that make it hard not to get lost. In the intricate hallways I ask a couple of people in white coats for directions and finally locate the laboratory at the far end of a small concrete patio. I push open the heavy and loudly squeaking door to what is to become one of my main fieldwork sites.*
>
> *When I enter I find myself in a large, slightly dark room, with small, partly blinded windows that I later find out cannot be opened, resulting in a rather thick atmosphere, in spite of the air-conditioning constantly humming in the background. The fluorescent lamps hanging from the ceiling are supplemented by the flickering and changing lights and colours displayed on the computer screens that are switched on. They are placed at the centre of the room on three long double rows of tables, where nine younger researchers are sitting looking at their individual screens. All of them are wearing headphones, listening to music while they work. Except for the sounds of keyboards and the clicking of 'mice', the silence in the room is striking. Nobody looks up as I enter – eyes fixed on the screens. I say hallo, and finally a young man gets up, smiles and says: 'You must be Kristina'.*

Computers constitute the main tool for the researchers in the laboratory, most of whom have a background in electrical, computer or other forms of engineering. The computers contain databases with images, sounds and ongoing projects, as well as archives, different types of program and code-writing software, academic papers and reports. A locker placed up against one of the white walls contains the usual 'computer stuff', such as cables in different sizes and dimensions, spare parts, printer paper, drivers and program manuals, as well as materials specific to the biometric laboratory. The latter include sensors,[4] voice recorders, pulse measuring devices and so on, depending on the tasks. Researchers have added their personal touch to the locker doors, gluing on them a plethora of internal jokes, as well as funny and strange photos of current and former members of the changing research team who have worked in the laboratory over the years.

In the laboratory, researchers process body fragments and behavioural patterns for use in biometric systems. The work in this laboratory focuses on vein structures, face dimensions and fingerprinting – subjects that are also sometimes known as biological or 'hard' biometrics – as well as signature, use of the voice, ways of walking (gait) and heartbeat rhythm, also known as 'soft' or 'behavioural biometrics'.[5] The researchers spend their time exploring the aptness of these different body fragments, sounds and rhythms for biometric identification; they define ways of registering users and their biometric characteristics ('enrolment'); produce representations of biometric characteristics ('capturing'); and 'process' the digital traces of body parts (e.g. images of fingers and veins) by using different algorithmic approaches.[6] Through this work the research engineers engage in abstracting, translating and mapping and in experimenting with new body-fragments, algorithms, codes and filters, continuously striving to push the frontiers of biometric science.

However, laboratory researchers also spend time testing different biometric devices made by others, such as fingerprint sensors, and in developing privacy-enhancing technologies, such as template protection, anti-spoofing applications (e.g. algorithms that can detect fake from real faces), and imagining the possible ways in which biometric technologies can be circumvented. The researchers are thus engaged in crafting, ordering, fixing, experimenting with and, not least, imagining new technologies. And, like craftsmen, they 'make things' (cf. Harvey and Venkatesan 2010) – in this case algorithms, code and programs that will make it possible for biometric systems to automatically register and recognize particular individuals through the use of their fingers, faces, veins, eyes, signature, gait and so on.

For these biometric researchers, the main aim of their work and of the practical applications of the technologies they are developing is the creation of 'a seamless world', that is, a world in which the opening of smartphones, computers and, for example, vehicle settings (mirrors and seats) are automatically enabled by a finger; office doors are opened by a voice, home lighting, TV and radio programs are turned on by a particular type of gait/movement; and people with disabilities will be able to use their voice or face as ID instead of having to sign documents, among others. This is indeed a future of individually customized surroundings that enables a seamless *and* safe everyday life. Nevertheless, where there is 'seamlessness', there are also 'seams'. And wherever biometric technologies are implemented, they also establish some form of automatized boundary between those bodies that have access to particular locations, devices etc. and those that do not. While the work in the biometric labs I have visited is therefore rarely directed explicitly at improving what is conventionally thought of as 'border control', the installation of biometric gates and the use of other biometric devices in airports, at land borders and hotspots often depend on the technologies developed and enhanced in labs. Laboratory work and the software that come out of it thereby become an important part of the border world and of the infrastructure that is put in place to regulate migrant and other travellers' mobility at borders (cf. Xiao and Lindquist 2014; Larkin 2013; Anand 2017). As such, biometric laboratories constitute an important site in the border world.

Whereas an ethnographic approach to labs in the context of biometric technologies and migration may be rare, as noted in the general introduction to this book, ethnographic research into science laboratories, in general, has a long track record in the field of science and technology studies.

Laboratory studies

Laboratory studies have flourished since the late 1970s, particularly in sociology and science and technology studies, and scholars such as Knorr-Cetina (1981, 1995), Latour and Woolgar (Latour and Woolgar 1979/86; Latour 1983, 1987, 1999) and Traweek (1988) have had a prominent role in shaping this field of interest. In these studies, the 'laboratory' not only figures as a concrete physical site where researchers engage in everyday practices of science making but is also used as an analytical notion which focuses on how laboratory scientists produce scientific facts through their daily lab practices and negotiations. In laboratory work, elements of the 'natural world', such as plants and soil, are abstracted from their 'original environment' and translated into new types of objects through their relations to other materials, pieces of equipment, techniques of visualization, social practices and so on in and beyond the laboratory (Latour 1999). In biometric laboratories, these processes of distillation and of the production of scientific facts take off when research subjects or users are recruited, body fragments such as the wrist veins of these users are captured in various locations, and these are then transformed into digital images through infrared-capturing devices. These vein images are subsequently related and organized in a wrist-vein database from where the process of refining continues. Like other researchers in science laboratories, biometric researchers thereby translate objects into their scientific 'extractions' or 'purified versions' (Knorr-Cetina 1995; Latour 1983, 1999). This translation takes place through what Latour (1999) calls processes of 'inscription', that is, the production of visual representations of wrist veins, for example. 'Immutable mobiles' are examples of such inscriptions or representations that can be transported across time and space as singular fixed forms, like for instance graphs, scientific text and maps, regardless of the lively and changeable materials and arrangements from which they are generated (ibid.: 307).

Science studies inspired by Latour and others argue that the general notion of clear-cut natural and social scientific objects, which science laboratories also operate with, is based on a thinking that separates the natural, social and material into distinct ontological categories. Scientific and other objects, Latour holds, are symmetrical and entangled 'hybrid' phenomena that are constituted in networks of actors (things, humans) processes, symbols and information, among others. Latour exemplifies this entanglement in the following quote: 'Press the most innocent aerosol button and you'll be heading for the Antarctic, and from there to the University of California at Irvine, the mountain ranges of Lyon, the chemistry of inert gases, and then maybe to the United Nations, but this fragile thread will be broken into as many segments as there are pure disciplines' (Latour 1993: 2). Along similar lines, this part of the book, with its focus on the making of biometric technologies, takes us from science labs to biometric and border conferences, standardization forums, associational meetings and vendor fairs. Rather than attempting to find the social 'beneath' the scientific, the question then becomes, for instance, at which point in knowledge production the social, the material, science and politics are connected, and how. At a more prosaic level, the success of scientific inventions, and for our purposes also of individual biometric labs, also depends on the mobilization of particular types of (social, economic and political) interests and connections beyond the labs. Researchers need to bring the world to the lab explicitly, as well as take the lab to the world.[7]

The ethnographic approach in the following chapters is inspired by laboratory studies and the ways in which labs are seen as places of practical daily scientific work that translates messy 'natural' phenomena such as bodies into scientifically endorsed representations that are abstracted from their embeddedness in the world. Furthermore, the conceptualization of labs as nodes in networks that include different practices, places, actors, agendas and political and economic processes, rather than merely physically bounded scientific workplaces, is another source of inspiration. In other words, biometric technologies are not simply produced in research labs, and the work with the 'biological body' in biometric science is only possible after an arduous process of separation of the body from its multiple socio-material entanglements.

The site: a 'biometric community' in a biometric landscape

In the following chapters, I explore how the work of laboratory researchers is configured in relation to a network of mutually dependent people, places and practices by following some researchers' paths through 'the biometric community'. This community comprises a rather small, tight-knit and self-referential network congealing around a shared interest in biometric technologies. It is also a community where processes of what Ingold (2000) and others call 'enskilment' take place. Ingold uses the word *enskilment* to refer to a particular 'education of attention' and embodiment of information that is essential to practical conduct in a specific environment (Ingold 2000; cf. Palsson 1994; Grasseni 2010). Through their participation in a plethora of events and forums in the biometric landscape, young researchers are trained to be attentive to certain issues such as key actors, trending topics and funding interests, and to learn to embody information that is essential to practical conduct and navigation in this specific community. The 'biometric community' forms part of the larger border world assemblage by virtue of the political and economic interests that are mobilized in biometric technologies as security-producing and mobility-regulating infrastructure on borders, whether they be hotspots, airport border controls or Danish consulates performing DNA tests abroad. It is also an assemblage which, when seen from the labs, is riddled with competition over resources, secret projects and security restrictions.

So how to study such a site ethnographically?

Methods and positions

Having not worked with biometric technologies before, I decided to participate in a summer course on biometrics at a technical university. The course teacher turned out to be an internationally renowned, well-positioned, highly esteemed professor of biometrics who generously shared his limited time and kindly opened the doors to more contacts, thus facilitating my fieldwork options. From this moment, I was met with generosity and a willingness to share knowledge almost everywhere I went among researchers. Offhand, it made good sense for both biometric researchers and industry players to be interested in working with a social scientist. The 'people skills and knowledge' that are considered to be the speciality of social scientists in general and of anthropologists, in particular, are often highly valued in the biometric field. One good reason might be that all EU projects, including those in biometrics, need the affiliation of social scientists. However, the latter are sometimes considered a 'necessary nuisance' because they (or should I say 'we'?) are perceived to complicate things unduly and sometimes take ethical stances that make it more time-consuming and difficult to advance science. In large projects, social scientists often cover the ethical or data privacy aspects and/or are positioned as user experts with particular insights into user behaviour. Knowing trustworthy social scientists can therefore be of great benefit to biometric researchers, just as biometric researchers can be important to anthropologists looking for the next funding possibility or the next round of fieldwork. On the one hand, it might have been this scenario that made it relatively easy for me to obtain access and find a legitimate research position both in labs and at conferences. On the other hand, I also believe that my access was shaped by an actual interest in and curiosity about social science and the work of anthropologists, as well as the lab researchers' own ideals of scientific openness.

A 'people person' in biometrics

Having previously worked with 'user experience' for two small companies, the role of 'user expert' was not unfamiliar to me, and during fieldwork I did indeed provide input to discussions about such issues, using the best of my knowledge. However, the delegation of social and ethical concerns to the social scientist, while quite common, points to the way in which the natural, the technical and the social are often considered separate domains. In the context of biometrics and other natural, technical and some social sciences, 'the social' is most often conflated with user perspectives and behaviour, which only need to be handled the moment the biometric technologies and installations are put to use. The social in this context is constituted either as an obstacle or, as Law puts it, 'a more or less unfortunate *afterthought* that spoils – or at least might spoil – well engineered technical relations' (Law 2011: 6, author's emphasis) but which, if handled well, might facilitate the smooth operation of the technologies. Arguing, like Latour, for the need to view all relations as simultaneously natural, social and technological, Law argues: 'There *is* no technical without the social, except in the dreams of engineers – and even *those* dreams are social' (Law 2011: 6, author's emphasis). In this context, the position as a 'people's person' with expertise in user experience made my presence more pertinent. Furthermore, at times, and particularly in relation to evaluations of biometric technologies, this role added value to my perspectives and opinions. Even in this context, as already noted, it was made clear to me from the outset that there were projects and work processes in the labs that were classified and to which I therefore could not be given access. Mostly these were projects dealing with state security, which, at times, also involved organizations like the US Homeland Security and police forces.

Fieldwork in practice

In both labs, I spent my time observing the different activities that were taking place, participating in the work as much as possible (e.g. in user trials and database captures, as well as user-experience work). I would have coffees and lunch at the lab and sometimes meet up with the researchers after working hours. Rather than being solely for reasons of fieldwork, this was also simply because the lab researchers were interesting, nice and fun people to be around. I would often sit beside individual researchers when they were performing their daily tasks (usually) in front of the computer and enquire into their practices. I conducted formal interviews with most of the researchers and with the head of the two research labs. A variety of different events and settings outside the labs, most of which were directly related to the researchers' work, also became part of my field. Concretely they entailed participation in conferences on biometrics, border security and image processing, in a vendor fair, as well as in biometric interest-group meetings and interdisciplinary network meetings. As other scholars have pointed out, such events, conferences, fairs and network meetings are 'ideal social settings where professionals perform for peers, convey information, tell stories and "embody" their professional identities' (Baird 2017:3). These people, places and networks were all part of what was referred to as 'the biometric community'. The older researchers would generally circulate with ease and familiarity in this 'community', whereas the younger researchers had to become gradually enskilled and to get to know the topography of the particular social landscapes that constituted it.

The chapters

Chapter 1, *Body cartographers: mapping bodies and borders in the laboratory*, concentrates mainly on the everyday sociotechnical practices, tinkering and experimentation that makes laboratory work engaging to the biometric researchers and that enables the use of body fragments for purposes of biometric identification and recognition. By using the analogy of cartography and coining the term *body cartographers*, the chapter argues that researchers, rather like explorers and cartographers, explore new body landscapes, experimenting not least with alternative (automatized) approaches to map them. The resulting 'body maps' are conceived as being unique to particular individuals and are generated by, and used in, biometric border systems.

Chapter 2, *'The biometric community': friends, foes and the political economy of biometrics*, portrays the researchers' engagement in a different form of cartographic and exploratory endeavour. This chapter demonstrates the researchers' need to extend laboratory practices and make sense of the social, political and economic biometric landscape beyond the labs, mostly in order to acquire the knowledge and skills needed to navigate and discover the shortest paths to collaborations and the next sources of funding while striving to maintain their scientific integrity. The chapter also highlights how the dynamics of this biometric landscape co-configures laboratory work and, ultimately, the ways in which biometric technologies are configured. This is the chapter in which we get a feel for the stakes and the different connections and disconnections, rivalries, symbioses and adaptations that characterize the biometric landscape and inform the researchers' engagement with it.

The type of landscape at stake in Chapter 2 is different from the body land-scapes that the researchers map scientifically in the lab in Chapter 1, as is the process of mapping. Nevertheless, the two forms of mapping are intricately entwined and indispensable to one another, both being shaped by the War on Terror and the fear of illegal migration, and both constitute part of the border world.

In a sense, the two chapters also constitute a form of cartography of my own, an exploratory journey through different biometric landscapes constituted by multiple, partly overlapping sites that are patched together by my own gradual movement through them rather than by any internal coherence of perspectives. The chapters explore the slow, but rather overwhelming, details of laboratory work, as well as the quick pace of people, presentations and politics at biometric events of different types. In keeping with, and perhaps as a reflection of, this complex and sprawling landscape, in the following, I do not aim to pull all the threads together, nor do I provide a final or coherent tale of biometric labs in the border world. Instead, I hope to engage the reader in a journey through the biometric landscape as seen from the laboratory at a particular point in time and with an eye to complexity, hopefully providing an understanding of the aspirations and practices of biometric lab researchers, the making of biometric technologies and the biometric border world.

Notes

1 For the sake of variety, throughout the text I use both the term *laboratory* and the shorter *lab* to describe the workplaces of biometric researchers.
2 *Presentation attack* is the standardized term for the attempt to circumvent a sensor by presenting the sensor with an object such as a fake finger or face mask. However, in daily practice it is still the word *spoofing* that is used. Research dedicated to averting such 'attacks' against biometrics is also known formally as 'Presentation Attack Detection' (PAD) or more informally as 'anti-spoofing research'. See Grünenberg forthcoming.
3 For the sake of anonymity, and since most people who work on biometrics in Europe know one another, I have chosen to change all names and places.
4 Sensors are physical devices used to 'capture' or record and transform fingerprints, for example, into digital formats.
5 In the standardized biometric vocabulary, however, there is no clear line between biological and behavioural biometrics, since, for example, placing a finger on a sensor is also considered a behavioural practice. See https://christoph-busch.de/standards.html#370205.
6 Briefly speaking, an algorithm is an ideal mathematical recipe or model of how best to solve a problem. In biometric research and practice, different algorithms are used for different tasks during the processing of bodies for biometric systems. Algorithms cannot stand alone but need to be operationalized by mathematical code(s) and the scripture of different types of computer programs (cf. Dourish 2016).
7 See, for instance, Latour's 1987 example of the diesel engine and the 'pasteurization' of France (see also Latour 1999).

1 Body cartographers

Mapping bodies and borders in the laboratory

The cartographer's dream is that of a perfect map, the map that would perfectly represent a territory, a dream of divine knowledge: a map that has haunted the ideology of representation throughout history: a map so detailed that it coincides with real space.

– Vrbančić (2005: 313)

In this chapter, I move into the biometric lab and engage with the lab scientists and their daily practices. By using the work on wrist veins as an example, as well as the analogy of cartography, I argue that biometric researchers, like the inherent aspirations of traditional cartographers, strive to establish an exact fit between the territories of the body and the representations of it that they produce. Additionally, like the inquisitiveness of New World explorers, sometimes cartographers themselves who would bring back cartographic details of hitherto unknown or undescribed territories, biometric researchers are continuously exploring and experimenting with new body parts or features that may be enrolled as biometric characteristics and transformed into a form of map. Finally, I argue that, like the border-making effects of cartographic maps, the representations of the body that biometric researchers produce have a 'map-like effect' in that they produce borders through the body itself. When employed in the context of migration, such 'body maps' are used to register or verify the IDs of migrants and refugees, thus permitting or alternatively preventing their onward mobility.

By understanding the work of biometric researchers through the analogy of cartography, I aim to highlight both their wishes for scientific precision and the exploratory inquisitiveness that motivates and characterizes most of their research practices. I also stress, however, the researchers' acute awareness of the fallibility of the type of map-making that they enable, which is automatized in biometric systems and forms an important part of the biometric border world. The analogy between cartography and biometrics draws its strength from the ways in which landscape metaphors are used to conceptualize and describe biometric characteristics such as fingerprints in biometric teaching and textbooks, as well as by biometric researchers themselves.

The ID-technology lab – continued

The ID-tech lab is part of what is known as the 'the academic group for biometric technologies' at the technical university, which consists of an academic research group and a technology testing entity. Richard, the head of the academic group and the lab, is in his mid-40s and has a PhD in electrical engineering. He has worked in ID technologies since the 1990s and headed the academic group for biometrics at the technical university for almost two decades. At the ID-tech lab, the number of researchers, students and assistants present in the lab on a day-to-day basis varies between 6 and 12, depending on whether those of them who still study have to do course work at the university, whether they might be involved in evaluation activities, are attending conferences or are travelling. Apart from their research and processing work with biometric modalities, as mentioned in the introduction, the lab researchers also test different biometric devices such as fingerprint sensors, which are used to 'photograph' ('capture') and convert fingerprints into digital representations that can be stored for later verification of identity. At times, this requires the researchers to spend time at the clients' locations, often private companies. Like other labs in the field this lab struggles to find a balance between what is conceived as basic academic research that will seem legitimate to academic peers and what is considered applied research focusing more on practical issues, a struggle also described by other scholars (cf. Gorm-Hansen 2011; Hoffman 2017).[1] However, this particular ID-tech laboratory seems to have struck a pragmatic balance. As Richard put it, 'We let the external funding as well as the funding from private companies finance our research'. By setting aside part of the industry and other (particularly EU) funding for their own basic research interests, Richard tries to ensure that the lab is still at the forefront of academic science. Indeed, in daily lab practices, the distinction between basic and applied research is not clear-cut, and the laboratory researchers mostly work collaboratively on both academic and more practice-oriented tasks.

However, Richard spends most of his time ensuring the financing of the research team, seconded by two other more senior members of the research team: Julia, who has worked for a biometric company, and Rafaél, who regularly attends conferences and focuses on user perspectives. Having worked with biometrics since the 1990s, Richard also attracts clients because of the reputation for expertise and experience he has established.

The laboratory as a social space

For an outsider, the laboratory initially looks like a place of deep concentration on individual work processes and highly skilled coding and mathematical work (see also Cohn 2010). There is seemingly not a lot of exchange going on in the quiet room where everyone is placed in front of their individual computer. It is not until I am invited into the laboratory online chat forum that I realize the intense sociality and processes of enskilment that are taking place and that I suddenly understand why the silence in the laboratory is often punctuated with what initially seems like random laughter by one or other researcher. Several chat forums revolve around laboratory work. General laboratory chats also unify the researchers in the physical lab with other researchers placed two floors up in another building and with people previously or still partially affiliated to the laboratory, who are all involved in daily exchanges, including good-humoured jokes about biometrics and individual researchers (as well as occasionally about the 'anthropologist-spy' who is visiting the lab). The researchers also share a channel that focuses on social events, such as dinners, birthdays and volleyball matches in the nearby sports hall, as well as several forums involving exchanges between those who are involved in specific projects that are only accessible to the participating researchers. In these project-related channels, the progress of, the approaches to and the tasks of the projects are discussed among participating researchers. These chat communications are combined with the daily social routine of morning coffee and snacks in the cafeteria at 11 a.m., lunch in the canteen at 2 or 3 p.m., afternoon shopping for chocolate. During these daily routines, the atmosphere is generally good-humoured, and as the researchers all come from different provinces, and in the case of three of them from other countries, teasing remarks and discussions often revolve around linguistic and cultural issues jokingly attributed to different regions. Once a month, there is a laboratory meeting headed by Richard, in which tasks are prioritized and assigned to the researchers and the technical approaches and progress of different projects are evaluated.

Negotiating approaches, shaping practices and establishing authority

Apart from the daily social rituals, there is also the occasional removal of headphones in order to discuss, for example, ways of processing body images for biometrics: Which algorithms, codes and programs to employ in order to extract the important features from biometric samples? Which filters to apply to the images of biometric characteristics in order to modify or enhance the properties of the vein image? Why do certain errors occur? and Which types of visual renderings are best, meaning the most accurate, as well as being aesthetically pleasing to look at?

Elsa is one of the researchers who works with anti-spoofing and fingerprints. On one of my days in the lab, I see a sort of drawing on Elsa's screen. I find the pattern with blue lines criss-crossing in the shape of a fingertip interesting and nice and want to take a picture of it, but Elsa tells me not to: 'No, don't take a picture of this one . . . this was a test, and it came out very messy. This one is better'. She shows me what to my eyes is a much less interesting and more conventional line graph.

Graphs and other forms of visualization of laboratory work are not only persuasive and useful; they also help establish academic authority, as Latour (1987) has argued (cf. Richards 2003). For Elsa, graphs are also evidence of a 'tidy', organized and thus more legitimate piece of research. Nonetheless, the invention of novel, colourful and visually pleasing ways of representing the individual

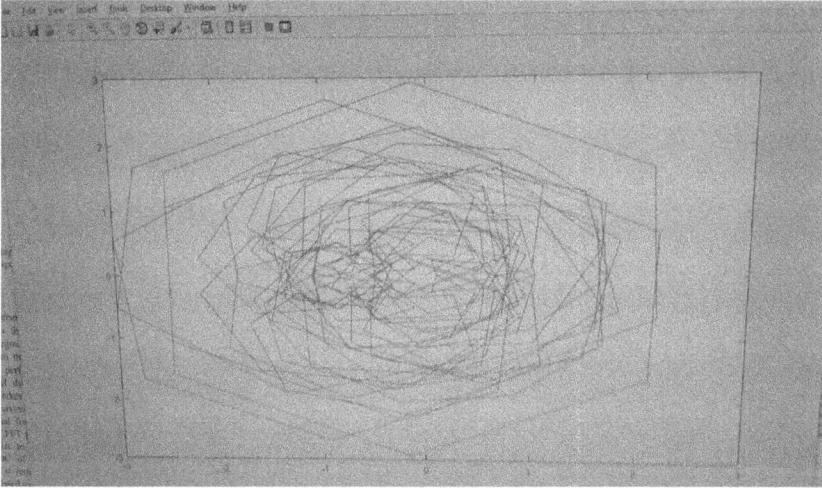

Figures 1.1 and *1.2* Two representations of fingerprint spoofing
Source: Photo by the author.

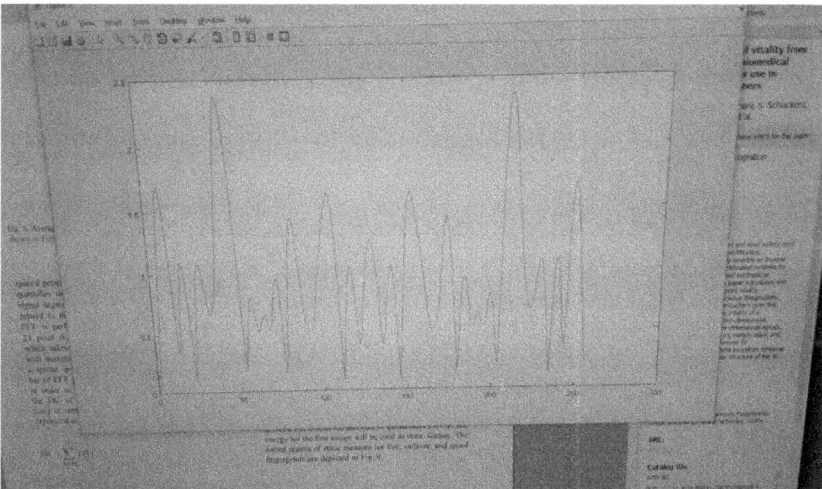

Figure 1.2

topography of fingers, faces and veins is also considered fun and an important source of creativity and satisfaction by the researchers. Which biometric characteristics the lab engages with and how, as well as how the results produced by the lab are visualized in, for example, academic publications then depends on a multitude of circumstances and 'small decisions' based on daily chats and live discussions of the particular tasks, the aesthetic and methodological preferences of the individual researchers and their scientific fields of competence. The graph produced by Elsa thus constitutes the culmination of the daily laboratory discussions and decisions about how to approach the processing of biometric characteristics and reduce the complexity of the digital body image. These practices and circumstances all shape, but are not visible, in the final research results into which they are, however, inscribed (Latour 1987). In the lab, results can come in the form of inventions of new or modifications of already-known algorithms, codes and programs for processing body fragments for biometrics, as well as in new visual renderings or inscriptions like Elsa's graph, which ultimately may end up in academic publications, reports and conference papers.

Like the multiple social dynamics that shape the research practices in the lab, it is also often a mixture of scientific interest, staff issues, funding opportunities and personal qualifications that decides which biometric technologies actually turn into research topics in the lab and how they are approached scientifically. Paola is a 25-year-old PhD student in the lab who has just initiated her work with wrist veins, or, to be more precise, with images of wrist veins. In her case, it was a particular combination of circumstances that shaped her topic of research. These circumstances included the fact that a previous researcher had left the research team for a better-paid and more stable position in a biometric company, thus leaving his work on vein biometrics half-finished. Furthermore, the laboratory has taken on a project on vein recognition in collaboration with a governmental institution in an Asian country. Paola was supposed to have continued the work of her former colleague on ways of capturing the veins by experimenting with capturing devices, such as different types of infrared camera. But she objected to this, explaining, 'I am not an electronically savvy person. I am better at image processing'. As a result, the project was reformulated, and instead, she was asked to do 'image processing', that is, processing the images of the wrist – veins in order to enable their use as biometric identifiers in automatized biometric systems. In her work, Paola experiments with different ways of approaching the veins. Like a cartographer, she registers and maps the topographical properties of the veins and thereby enacts a particular vein landscape.

From body part to image and image to map: working with veins

The use of landscape and cartographic metaphors in research on biometrics first came to my attention when I saw a video made by a well-known biometric researcher who specializes in fingerprint biometrics. The video, which was used in a university course on biometrics I attended early on in my fieldwork, features a small plane overflying a fingertip depicted as a pattern of valleys and ridges and dropping little yellow minutiae points where the patterns are defined as being particular to the specific fingertip (see also the general introduction on technologies). Historically cartographic map-making, among other things, served to 'open up' and facilitate travel through and entry into 'unknown' territories for explorers (who were sometimes cartographers themselves), who were often followed by representatives of states or empires, churches, the military and others interested in domesticating 'terra incognita'. As geographer Kenneth Olwig notes, cartography made it possible to 'take a complex mosaic' of land attuned to particular ecologies and 'reduce it to geometric, privately owned fields' (K.R. Olwig 2004: 10). Olwig further argues that It was cartography that created 'both the idea of a sharp unbroken skin-like bounding of national territories within the striated Euclidean space of the map, but also the idea that the Euclidean space encircled by these boundaries was of a uniform, homogenous character' (K.R. Olwig, personal communication 04.03.19). The cartographic descriptions, mapping and consequent reduction of territories in this context frequently served to demarcate spatial territorial borders, as well as the political boundaries of a piece of land, in spite of already-available local understandings of the area, and thus enable the control of passage. In a similar way, biometric maps permit the abstraction, reduction and opening up of the individual body for topographical inspection and in this case for the automated verification of the correlation between a stored, mathematically produced body map (or template) and a body territory (the live-scanned body, i.e. the finger, face, vein etc.) conceived of as naturally bounded and homogenous.

Vascular biometrics

Finger, palm, hand and retinal veins constitute parts of the body that have recently been made subjects of biometric mapping. This biometric modality has attracted quite a lot of research interest in recent years, as it is considered to be one of the safest and most 'spoof-proof' of technologies and there are several unexplored avenues in the field. According to biometric research the structure and patterns of the veins are unique to individuals, even among twins, who may pose problems for other biometric technologies. In addition, the veins are, of course, located under the skin and are not readily visible without infrared technology. The process also requires the presence of blood flow, which makes them hard, but not impossible, to forge (see the Epilogue). Currently sensors using finger veins, eyeballs/ retina veins, palm veins, veins on the back of the hand and wrist veins are either already in production or are being worked on.[2] The comparison of veins for recognition or identification can take place in different ways. One of the conventional ways is by establishing and comparing what is known as 'minutiae points', that is, points located in particular places in individual body maps or patterns. These minutiae draw on landscape and topographical metaphors that were originally established in relation to fingerprints, with labels such as ridges, valleys, islands, bifurcations and endings in order to define the particularity of individual fingerprint landscapes. The use of minutiae is also prominent in vein biometrics where the minutiae serve to make individual maps of the specific vein 'landscape' and ultimately distinguish one set of veins from another.

But let us return to our PhD student Paola in the lab, who has to find or construct a wrist-vein database before she can start her image-processing work.[3]

Building a database and imagining 'the wild'

The first and most important part of the process of working with wrist veins is to find or create a database, Paola explains, in her case one consisting of images of 'wrist veins'.

A biometric database is basically a storage and classification system consisting of what are known as 'pre- and post-processed' stored digital images (of, e.g., fingerprints, veins, faces), frequencies (e.g. of sound/gait) or traces produced from other biometric characteristics. However, finding a database that is accessible for use is often a problem. Paola tells me: 'There are good vein databases out there, but the companies that have made them do not want to share them'. Databases are very expensive to build, and it is something for which it is hard to obtain research funding. Commercial companies have means of their own, but, according to Paola, they are not interested in sharing their methodologies and information, which, in a competitive environment, is seen as producing market advantages and potential patents.

Constructing and assembling a database, then, is no small feat, and databases are intricately connected to the political economy of biometrics. They also require lengthy procedures. First, the researchers need to find and recruit people who are prepared to put their wrists and veins at the disposal of a biometric science lab. In the ID-tech lab, apart of the participation of the researchers themselves, this is usually achieved by giving free cinema tickets or small sums of money to students, staff, friends and others in the networks of the researchers, preferably people of various ages and both men and women. The visibility and imaging quality of the veins change depending on the bodies of the users. Therefore, researchers dealing

Figure 1.3 Image from vein database

Source: Courtesy of the laboratory.

with this topic generally need to experiment with variations, such as the depths of the veins, the thickness of the skin, the presence of hair, moles and scars, and the particular circumstances of the setting, such as the temperature, humidity and lighting in which the veins are captured. In this context, then, researchers have to imagine different 'real-life scenarios' or applications in 'the wild' that might have an effect on the readability of the veins from the database. This leads to experiments with capturing processes under different circumstances, for example, giving users different tasks, such as carrying heavy loads, putting their hands in cold water and squeezing something before capturing the veins in order to see what difference it makes in reading veins (cf. Yuksel et al. 2011). Then, there is the process of registering the users and capturing their veins. This implies a new set of materials and practices, such as a user-friendly sensor with infrared light, computer programs, numbering systems to register and anonymize the users, and so forth, enskilling users so they may interact with the sensors in an appropriate fashion, and so on.

Databases are nonetheless essential and constitute the basic material for the development of biometric technologies. Without good databases, there is no material from which to develop the technologies. The different types of algorithm are trained and tested, the codes are written and the programs are made with, and in relation to, particular sets of vein images from the database. If the database is somehow badly constructed (bad-quality images; varied positions; inferior capturing devices; faulty registration of images; gender-race- or age-biased database subjects) or too small and the algorithm has therefore not been trained adequately, Paola told me: 'Your whole material is skewed, or the system will never work *in the wild*'. In that case, the bias of the database would have been installed into the system. In this context, the system might only work with the database with which it has been trained and not be able to detect other veins in a way that makes them usable for the verification of a person's ID, or else it may falsely reject or accept the wrong people.

The wild is a contrastive term used by the researchers that constitutes the laboratory as a particular type of 'controlled space' that allows for extensive experimentation. This conceptualization also denotes the demarcation between the controllable environment inside the laboratory, where body parts are enrolled, transformed, mapped and 'domesticated', and the uncontrollable conditions 'outside the laboratory'. Here, bodies 'run wild', and environmental factors such as those described earlier may make it impossible to actually use biometric sensors and installations unless some of the laboratory's practices and conditions, such as sufficient and stable sources of light and temperature, compatible software and sufficient user guidance, are extended into such 'wild, real life' settings (cf. Latour 1983).

Apart from the vein database built in the laboratory, which she uses to train the algorithms, Paola uses a publicly accessible database from Singapore to test it. The Singapore database contains wrist-vein images taken in Singapore between 2009 and 2011 from 131 people from different countries. The images are captured with near-infrared cameras, which makes it possible to view patterns under the skin. In other words, the wrist vein image database was generated by capturing a small fraction of the bodies of different people, at different times, storing them on a server and enabling them to be used by researchers across the world. The databases used by Paola then constitute repositories of different bodies, times and spaces that coexist in a particular laboratory time-space, which differs from the way in which times and spaces are usually structured linearly in 'the wild'. The veins have become mobile in this first instance as digital images assembled in a database, and as such, they can travel independently of the 131 bodies to which they are normally attached.

Sorting out the noise

Paola's task is to experiment with ways of pre-processing the wrist-vein images in the database. Paola goes through images in the database one by one to detect the variations in vein position. She uses her vision, which has been trained by the literature, discussions with colleagues and repeating practices, to analyse every vein image in order to find the best way to isolate what is known as the 'region of interest' (ROI). The ROI denotes the particular part of the vein topography that can be used for purposes of biometric identification. In this process, Paola defines what constitutes 'foreground' and 'background', marks the 'essentials' and sorts out the 'noise' and, thereby, defines her object of interest in each image. One of her fingers slides across the screen as she shows me what she has defined as the ROI of the wrist image. In this case, it constitutes the ways in which the veins branch, fork, bifurcate and end, as well as the points that can be used to generate a map from the specific configuration of the veins in this particular wrist.

In the first instance of experimentation, Paola manually filters out the parts of the images that she defines as not containing any clear vein structures. However, her aim is to write, train and test a new algorithm which can automatically detect the vein topography of any vein image. Instead of Paola or other researchers having to go through countless numbers of vein images, the algorithm is taught to 'see' and process the veins so that the image may be transformed into a body map or template at a later stage in the process. The whole point of biometric systems is, of course, that they can automatically generate and compare such (body) maps. The minute practices of preprocessing that Paola performs merely constitute the second step in the transformation from body part to image, to body map or template. However, the nature of this transformation is essential for the steps that follow, and different approaches can, according to Paola, ultimately mark the difference between secure or insecure identification of the individuals that have been enrolled into the biometric system. By the time the vein images have been registered, stored and are ready for comparison, they should constitute clearly marked vein patterns, beautiful patterns branching their way across Paola's screen.

Alternatively from here the minutiae points that are located at the pattern endings, ridges and branches can be calculated and turned into minutiae maps (Figure 1.2) and compared to points in other vein landscapes.

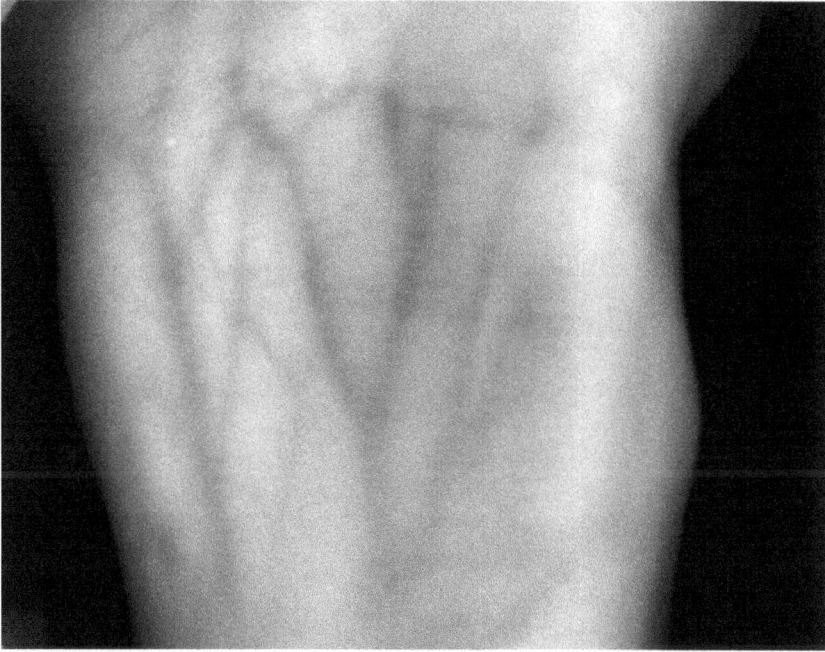

Figure 1.4 Image of wrist-vein processing
Source: Photo by the author.

Body cartographers and the not always so perfect maps

Although at several of these stages Paola no longer works with a living, throbbing wrist, but rather with maps of different orders of abstraction, such maps, as Møhl (2012) and others remind us, retain a form of relationship to their objects (here the body), as well as becoming objects that have an effect in the world (Møhl 2012: 76). Ultimately, it is the type of knowledge that Paola and her colleagues produce that makes it possible to generate the maps or templates used to stop migrants, such as Mukhtaar (see Simonsen, Chapter 5) or the relatives of Hanad (see Olwig, Chapter 7) in their onward journeys through the already-registered, individualized and fixed maps of their fingerprints or of the particular mapping of their DNA. Like cartographic maps, the biometric body maps then become a way of lending order to the world, and the seamless relationship between the map and the individual body landscape, potentially made possible by laboratory work, is indispensable to their success. As such, though the body is abstracted and mediated through the multiple practices of biometric research, the individual body, in the form its vein patterns or fingerprints, is still very much present in the map, and biometric maps have real effects. Body maps and templates produced in the lab thus enfold

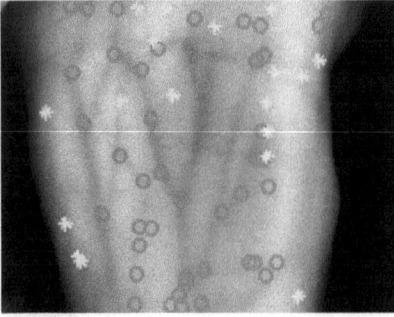

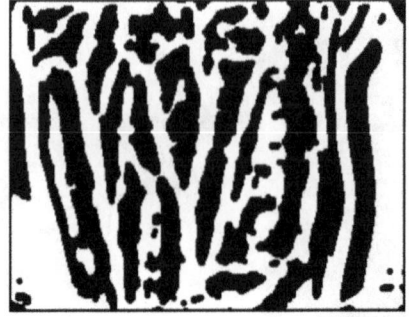

Figures 1.5–1.8 Four images of veins in different processing stages

Figure 1.6

Source: © 2011 IEEE [2]

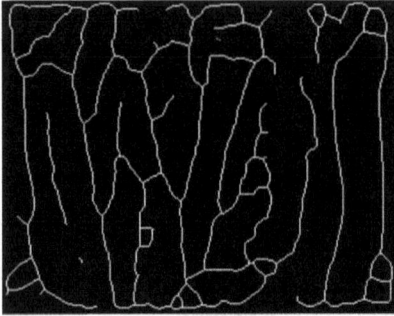

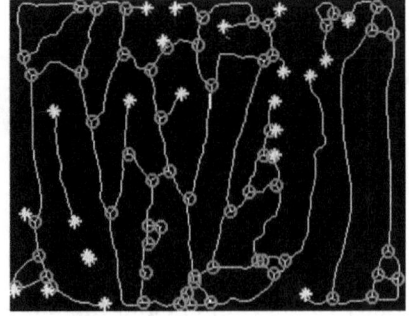

Figure 1.7

Figure 1.8

practices resonating with historically mediated cartographic visions of the world that imply the transformation of landscapes and territories into fixed, metrified and mathematically calculated topographies. Through the practices in the lab, and even more so in the implementation of biometric installations, the body, like the territory in cartography, is enacted as a relatively stable, unique, metrifiable and mappable entity untouched by its relational, temporal and spatial entanglements. Historically, cartographic practices of map-making have also been associated with 'the body politic' and political and ideological visions and representations of the state, of state borders and of territorial power (K.R. Olwig 2002; Cregan 2007; Kitchin and Dodge 2011; Møhl 2012; Lammes 2017). This metrification of the 'body politic' could, according to Kenneth Olwig, be considered central to the perceptual foundations of the biometric technologies of the present-day border(ed) world (K.R. Olwig, personal communication 04.03.2019; cf. Cregan 2007). Indeed, biometric maps, when used in the context of migration, co-constitute the borders of the state by separating those who are granted the right to enter national state territories and those who are not, based on maps of their bodies.

Because of their way of exploring and mapping the topographies of body landscapes, I have come to think of biometric research engineers as a form of 'body cartographer'. Like cartographers, these researchers understand the process of map-making as objective and as based on scientific facts, and they strive to make a correlation between body landscapes and their maps. The 'body maps' in the labs, like cartographic maps, emerge through playful 'iterative and citational practices', as well as aesthetic sensibilities (Kitchin and Dodge 2011: 111; K.R. Olwig 2018). Researchers are quite aware, however, that the maps are not the *whole story* of the individuals they come to represent. Separating the body from a form of inner substance, Peter, a professor of biometrics, expressed it as follows:

> The body is the outer appearance of the individual. What an individual is, and what identity is, that is hard to define! But the body is the tangible, observable outer surface of individuality – and the body as such is unique. It is a form of container of individuality and it like . . . eeh . . . well this is a weak analogy – no one ever asked me this before, so my answer is maybe a bit immature, but it's like if you want to identify a car or assign a car to the car owner, then you can go and read the serial number, the license plate or determine by colour, length or model. Somehow, from these observations you can make conclusions about the uniqueness of this car. Through the same analogy you can say that you can look at my eyes, or you can record my voice or you can look at my fingertips, and these are all identifiable patterns that allow you to identify my individuality, my individuum or singularity.

Although in the case of vein biometrics it is actually not just the outer surface that is captured, the distinction between the body as a somatic surface and the body as a form of inner self and principles that Peter makes resonates with the classical Platonic distinction between body and soul that is so prominent in contemporary western thought. He continued:

> What, then, is my identity . . .? Mmmm . . . that's difficult. Is it just this container that holds identity attributes? I think in the common understanding of how we use the term identity – it's more than that! Identity is somehow my spirit, my sensitive information, what I believe, what my principles are and so forth.

In Peter's perspective, this domain of sensitive information is not the working ground of biometrics but represents information that should be under the definite control of the individual.[4]

At the same time, the researchers are acutely aware of the innumerable things that may not work or that can go wrong in the process of map generation and reading, be it their own or those of the chosen algorithms. They also know about all the little detailed practices of algorithmic processing (preprocessing algorithms, feature extractions algorithms, matching algorithms, etc.) and their translation and implementation into larger assemblages of software, hardware, people and places which have to work together in order for the mapping practices to actually succeed in producing functional maps of the body and ultimately of the IDs of particular individuals. The mapping practices undertaken by biometric researchers also have to be continually worked upon. Bodies and body attributes change over time, like the territories mapped by cartographers. As Colin Turnbull argues, instances of 'mapping [. . .] are unstable acts of bricolage: the homogenisation of messy motleys that have to be continuously renewed and stabilized' (Turnbull 2010: 126). At times, bodies are even *made* to change through, for example, self-harming body modifications, such as when a young Somali man, in Chapter 6 by Simonsen, burns away his fingerprints to avoid his fingertips being mapped at the Italian border, a map that will end up in the EURODAC database and limit his possibilities for onward mobility. Furthermore, the concern with spoofers that may cheat biometric sensors with forged body parts, such as fake rubber fingers, masks, fake iris contact lenses and so on, constitutes a form of mess that requires a constant renewal of biometric research practices and keeps researchers on their toes (Grünenberg forthcoming).

Enskilling bodies for biometrics

In the ID-tech laboratory, as already mentioned, the researchers also work with evaluations of different types of sensors. Therefore, they also know of some of the potential problems that may occur in the subsequent phases of the application of biometrics and some of the ways in which it requires a particular form of body enskilment in order for the systems to work, as we shall see in the following example.

Today is the day that the evaluation of fingerprint sensors should start. The laboratory researchers are to test three different fingerprint sensors in order to inspect the ways in which they work on a number of parameters defined by a standardized procedure that contains a series of user-related and technical issues. The research crew has been busy for weeks recruiting users and instructing operators, that is, those who will help the users perform the task of testing the sensors. Olga, a young research assistant in the lab, has been particularly busy preparing the system, that is, the computer program that will connect the sensors to the computer and make sure that the fingerprint images and the non-identifiable user data are stored in the newly constructed database. However, she has had multiple problems with the communication between the sensors and the program and spent most of her afternoons and even some nights in the lab during this period trying to work out what was going on and why the sensors and the program she has carefully developed do not communicate. A couple of times, just when she thought it had started working and she had left the lab for the others to continue the tests, she was called back because it wasn't working anyway. Fingers were put on the sensor – to no avail. 'I really don't know what is going on; it was working when I tested it', she said as she got up briskly from behind the computer screen. She strode between the sensors and the courtyard, where she smoked nervously. Some students had already arrived at the lab in order to be 'enrolled', responding to the offer of cinema tickets for finger enrolments. In total, 800 people were to be enrolled over the following months with four prints of six fingers captured from three different sensors. After several days of troubleshooting, calling the sensor company to have it check the sensor software and checking the program components to see whether all the codes were working as they should, it turned out that it wasn't Olga's program which was the problem but that the different sensors, for unexplainable reasons, only worked when used in a certain order. Adding to this problem, one of the sensors persistently overheated and broke down, so Olga and her colleagues had to plug it in – and out – after capturing a few 'prints' in order to ensure that it continued working. 'Really, really strange', Olga burst out, when she finally located the problems. 'But these types of unexplainable things always happen'. Added to this, however, were other types of problems that materialized once the user tests started.

Thomas, who is working on his MA thesis on heart biometrics, is the first person to test the sensors. Sitting in front of the computer, he is instructed to put six of his fingers on each sensor four times. After considerable frustration and failed attempts to produce images from Thomas's finger, it turns out that each sensor requires a different type of pressure and a particular positioning of the finger in order to work. Not only does Thomas need to remember which finger to place where and when, but he also has to learn the particular bodily skills required to have his fingers 'captured' by each of the three sensors. The particular sensors in this respect do not simply 'capture' Thomas' fingerprints; they also format Thomas's body in particular ways through a process of bodily enskilment.

Figure 1.9 Image of fingerprint testing

Source: Photo by the author.

Even though the biometric researchers are quite aware of the potential challenges involved in the different steps of biometric processing, the aspirations of the 'body-cartographers', like those of the cartographers described by Vrbančić (2005) earlier, are still to close the gaps between body maps and body parts and thus provide accurate identity matches. Although, in principle, there is, in the researchers' understanding, a one-to-one relationship between the (living) body and its various forms of biometric representation, it is, they say, not possible to generate a 100% map with the technologies so far developed. In fact, this is one of the aspects of biometric research that researchers appreciate. As Peter, the professor of biometrics, put it when he was asked what it is about biometrics that motivates him, 'My general motivation is to tackle hard and yet unsolved problems, and in this regard it is interesting to work on biometrics, because it is always unsolved! (laughs); it works to a certain extent but you can never reach a 100 percent.'

From changing territories to seamless maps: fallible technologies and seamless security

The continuous processes of trial and error that take place in the labs in order to improve the correlation between maps and bodies are visible in the biometric research literature. Such documentation of extensive and rigorous experimentation, fault finding and errors also serves to testify to the academic rigour, academic legitimacy and credibility of laboratory science (Latour 1983). Nonetheless, the potential and actual errors or weaknesses displayed in the biometric research literature generally seem to disappear from view when biometric technologies are marketed, especially by large companies, to, for example, government and other agencies for the purposes of controlling border crossings or other types of access.

In these contexts, the biometric maps and templates and the living bodies become one and the same, the maps potentially being seen to speak the truth of the living bodies, just like landscapes and their cartographic representations are sometimes conflated and confounded (K.R. Olwig 2004). In this context, biometric technologies and the body maps and templates that serve as identifiers are also turned into (objective and) waterproof tools of security and the regulation of mobility and immobility across nation states. In an article on mapping, Holman Jones and Harris (2016) argue that maps 'mark intensities, layers of being – of bodies, of times, places and actions' (Holman Jones and Harris 2016: 5). The outcome of automatized biometric map-reading has a somewhat similar effect to this type of map reading in the context of the border world. In principle, and when the systems manage to work together, they enable particular authorities to tie different times, places and actions to individual bodies through the trajectories of their body maps and thus assess their legitimacy. When biometric 'body cartographers' map bodies, borders are potentially drawn, as in conventional cartography, but right there in the body – by a wrist, a finger, a face and so on, even when, as mentioned previously, the mapping strategies produced in the lab are not necessarily directed at migration control or borders. Such mapping practices also represent just one step in a larger assemblage of people, materials, sociotechnical arrangements, practices and 'little decisions' that produce biometric border technologies and co-configure the border world.

The biometric border world and the desire for the future

As mentioned in the chapter's introduction, the connections among the researchers' work on algorithms, the associated coding and programming and the border world are most noticeable in the biometric gates or corridors or at other types of border crossing throughout Europe.[5] These installations resonate with what Brian Larkin defines as infrastructure: 'built networks that facilitate the flow of goods, people, or ideas and allow for their exchange over space [. . .] They comprise the architecture for circulation' (Larkin 2013: 338) and 'enable (and/or, I would add, *disable*; cf. Anand 2017) the movement of "matter"' (Larkin 2013: 329, author's parentheses) – in this case, the movement of human matter. Drawing on Walter Benjamin, Larkin argues that infrastructures are more than technical objects as they also 'operate on the level of fantasy and desire [. . .] encode dreams of individuals and societies and are the vehicles whereby those fantasies are transmitted and made emotionally real' (ibid.: 333). These fantasies, according to Larkin, also work as 'filters through which the object is seen' (ibid.). Biometric technologies and systems have a form of infrastructural quality that corresponds to these descriptions. Like infrastructures, biometric technologies, in whatever shape or form, tend to hold the imagination and hopes of high-tech futures in a firm grip: 'the sense of awe and fascination they stimulate is an important part of their political effect'(Larkin 2013: 334; cf. Pedersen 2011). In the context of biometrics, this is a future tied in with notions of security. The fascination with the future is also linked with what Louise Amoore refers to as 'the politics of possibility', implying a new way of conceptualizing security. In her book of the same name, she refers to US journalist Ron Suskind's documentation of the post-9/11 Bush administration's 'one-percent doctrine' approach to the War on Terror. The doctrine encourages work with potentially exceptional occurrences and imaginable future scenarios in the present. Suskind quotes then vice president Dick Cheney as saying, 'Even if there is just a one percent chance of the unimaginable coming due, act as if it is certainty' (Suskind 2006 in Amoore 2013: 11). In the labs, such exceptional and potential future scenarios are addressed in, f or instance, the daily work on 'spoofing' and 'template protection', which involves imagining potential future attacks against biometric sensors. Like the work on 'real-life scenarios' in 'the wild' described earlier, this requires the researchers to engage with another type of 'wild card' and, in this case, to apply their imagination in order to avert potential future attacks against biometrics. What is, in effect, only a potential and imaginary threat thus becomes part of and is materialized in the daily lab work. The desire for a safe future and imaginaries of current and future efficient tech solutions inscribed in biometric technologies seem to be mediated, among other things, by the extended use of particular types of imagery, as well as pop-cultural, literary and filmic references to science-fiction universes. The use of such imagery is widespread in biometric newsletters, in biometric courses, at biometric and security conferences and when researchers are recruited for the labs or new research results are communicated through different types of media platform.

Imaginaries drawing on such sci-fi universes (see Figure 1.10) exist in parallel to the research literature and project reports, which generally include more prosaic images of or references to biometric fingerprints and sensors (see Figure 1.11).

It is, to a large extent, the current political focus on migration and terror and the financial investments in security, as well as the perceived (not the least symbolic) potential of biometric technologies in border control and other security domains, that shapes funding opportunities for researchers. However, the labs are not isolated, self-contained entities of science, research and technological development, where discoveries are simply made and then brought to the public (Latour 1983, 1999). Nor is their work merely a reflection of particular funding structures and trends. The labs themselves also play an important role in forging an interest in the future of biometric research, as well as a general interest in biometric technologies – two perspectives that go together.

By working on the seamless correspondence between e.g. sets of veins or fingerprints and their maps they become the co-producers of a particular notion of security that is perceived as being rooted in the body and is therefore held to be objective and true (Cf. Van der Ploeg 1999; Aas 2006; Magnet 2011; Gates 2011; Amoore 2013). More important, perhaps, by participating in projects that address the relationship between security and biometric technologies at biometric and other conferences and that focus on migration control and terror, among other topics, the researchers contribute to pushing the biometric agenda forward.

While ethical questions about biometric use and supposed infallibility do not figure prominently in their everyday work, partly because in the hierarchical

Figure 1.10 Image of fingerprint

Source: Creative commons.

Figure 1.11 Image of fingerprint

Source: Creative commons.

organization of labs, such questions would be handled by the head of the laboratory; when asked Piero, one of the researchers told me:

> You ask any electrical engineer and they will say – yes, of course any system
> has failures! For some reason people believe that technology always works
> and does not make mistakes and gives too much power to biometrics or any
> kind of system. This can become a way of avoiding responsibility! We should
> generally think about the consequences of these technologies ethically, politi-
> cally and even medically. But when I am developing a biometric tool in the
> laboratory I can still always take a step back – people on ethical committees
> elsewhere would let me know if it was wrong! [ethically questionable]. As an
> engineer I have to be able to take a step back and modify the tool accordingly.
> But there is no reason to involve such ethical committees at the level of basic
> research, when the technology is not yet working nor about to be applied.

While clearly seeing the potential pitfalls of technological hypes and the need
for general ethical discussions, Piero trusts in other people to determine whether
the tools that he is developing become problematic. In this way, he also turns the
laboratory into a safe space, a bubble of sorts, where free experiments can and
should be conducted. As we shall see, more clearly in the next chapter the laboratory however, is not much of a 'safe space', and the researchers need constantly
to work actively on building and mobilizing particular alliances in a biometric
social landscape that is furnished by biometric vendors guarding proprietor databases and algorithms as trading secrets, systems assembled from parts that do not
readily communicate and are owned by different stakeholders, heated discussions
about how these systems should work and not work, and local and international
rivalries.

   This chapter has focused on the research practices, aspirations and interests
of the researchers, as well as on how the work they are performing results in
a type of map that, like the maps of cartographers, are the result of opening up
and abstracting 'unknown' body territories, giving them names and calling them
their own. When applied in a migration context such maps turn the body into a
border landscape. In this context, the researchers and their biometric technologies
become enrolled in political agendas and dreams of a future in which technologies
are seen as the solution to issues of security, including illegal migration.

   In the following chapter, we see the researchers engaging in a quite different
form of cartographic and exploratory endeavour. Here, rather than body landscapes, they need to make sense of a social, political and economic biometric
landscape beyond the labs, one that is influenced by the War on Terror and the fear
of illegal migration. As such, it is inherently related to the border world.

Notes

1 In practice, it is questionable if there ever was such a thing as 'pure' basic research or
  clear-cut boundaries between 'basic research' and 'applied research'. As other scholars

have pointed out, research is always part of a social, political and economic environment (see Gorm-Hansen 2011 for interesting perspectives on this entanglement).

2 Haemoglobin in the blood contains oxygen when it is transported from the lungs to the tissues in the body by the arteries. When the blood flows back to the heart, this oxygen has been released. Deoxidized haemoglobin absorbs infrared light, making the vein patterns visible as dark lines.

3 Databases in the laboratories are also used to establish what is known as the 'ground truth'. By training algorithms on a manually configured vein data set and testing the algorithms on a different data set in the database, the 'ground truth' of the algorithm is established. From here, all other operations depart (see also Jaton 2017 for a more detailed discussion of this subject).

4 However, other lines of research work with automatized facial and body *expression analysis*, as Kelly Gates (2011) has shown. This type of analysis is oriented towards 'emotion detection' such as the detection of bad intentions in security settings or the detection of pain during operations (Gates 2011).

5 Biometric systems have been installed as border technologies in many other countries by now, among them the US, Ghana, Gambia, Canada, Hong Kong, Mexico, Australia, Korea and Japan.

2   The 'biometric community'

Friends, foes and the political economy of biometric technologies

Several pathways lead from the laboratory to the places and people, the organizations and committees, that all form a part of the biometric landscape. This is a vast and sprawling landscape of changing topography that is difficult to grasp in its complexity, with its manifold places and sites, institutions, people and networks. This is also the messy motley world encountered by the young researchers who are exploring and familiarizing themselves with it. Some sites, however, figure as central nodal points of interaction and for the more experienced researchers constitute part of 'the biometric community'. Through a series of empirical accounts that I call 'pathways', that is, pathways from the lab through the biometric landscape, in what follows I focus on four such sites and the ongoing dynamics, topics and sources of controversy that they encompass.

Pathway 1. 'Defining the wild and invasive species' describes the internal fractioning of 'the biometric community' following the establishment of biometrics as a particular field of interest for policy-makers, national security agencies and industry, and the consequent proliferation of actors wanting stakes in it. In this context, the pathway takes researchers into the field of biometric associations that are shaped by both personal relationships and senses of community, as well as by rivalry that it is hard not to become entangled in.

Pathway 2. 'Negotiating biometric border control' takes us to a biometric workshop at a border security conference and describes some of the controversies over the use made of biometric technologies at national borders. It draws up some of the marked positions in the landscape through the exchange among an EU applied researcher, a biometric expert and an industry player, who position each other as profit-driven technophiles and cautious technophobes respectively.

Pathway 3. 'Negotiating, standardizing and realizing biometrics' depicts the difficult act of achieving consensus in a standardization meeting of stakeholders with very different stakes and interests in biometric technologies, and how 'small decisions' in this forum may potentially change some of how European borders are arranged, industry players' market shares, and the work of laboratory researchers.

Pathway 4. 'Unveiling and concealing' draws attention to a dilemma that is inherent in the biometric field between the needs to not only unveil but also partially conceal new forms of biometrics and their respective advantages. Vendor fairs are one of the places where this dilemma stands out. Here the need to strike a balance between, on the one hand, the desire to present new and recently invented biometric systems and attract prospective buyers with a potential interest in biometrics and, on the other hand, the importance of not 'unveiling' too much in a landscape permeated by fierce competition and the need to patent new developments quickly becomes crucial.

I argue that, just as these four pathways through the biometric landscape lead researchers away from the physical labs, so, too, do they lead researchers back, with implications for the types of work and the technologies that are developed there. Biometric researchers therefore need to learn how to navigate in this landscape that also forms part of the border world.

When attending events at these sites, it quickly becomes clear that most of the participants know each other, often from other, similar events or from earlier collaborations. Furthermore, it becomes clear that several participants have changed positions over the years or simply had several roles simultaneously, for example, as former Federal Bureau of Investigation employees or a border guard commander now in biometric companies, as former biometric company employees turned national security officials, as researchers turned biometric vendors and as vendors turned researchers. In this context, it seems important initially to address the entanglement of economics, politics and academic aspirations to be on the frontline of research and of the future I described in Chapter 1.

9/11 and the proliferation of biometrics as security technologies

Whereas biometric technology research was for years, according to the researchers, an activity that drew no great attention and consisted of a few small research groups, the biometric landscape proliferated into a large and notoriously competitive field after the huge economic investments in these technologies as a consequence of the terrorist attacks on the Twin Towers in New York on 9/11. According to Gates, this development was spearheaded by the US, where biometric vendors and consultants convinced the US government that, if facial recognition systems had been available, the terrorists performing the 9/11 attack would have been caught quickly (Gates 2011; cf. Amoore 2013: 4/21).

In an interview, Richard, the head of ID-technology lab, argued that 9/11 changed the biometric field completely:

> In 2000, we were around four researcher groups who were working on and interested in biometrics; some worked on voice, some on face and some on (finger)-prints, hand geometry and iris, but there really wasn't a lot of interest. We knew that iris worked well, that fingerprints worked really well, and that voice and face worked really badly – hand geometry didn't really work well either. Then the 11th of September happened and,

I suppose, because of internal lobbying in the United States [. . .], in 2002 the US declared biometrics to be the solution to the security problems, and facial recognition to be the best biometric technique, even though the few researchers who had worked intensively on the topic knew that facial recognition was not working very well. This [situation], however, raised expectations in the world of biometrics, and of course where there are money and expectations – suddenly a lot of people become interested and claim a field.

In this process, the researchers' small, but relatively well-defined, network exploded, and according to Richard, the established biometric researchers lost ground:

A lot of people started claiming that they were experts in biometrics – both companies and researcher groups, even if they didn't have any experience whatsoever in the field. In my opinion, honestly, this resulted in huge investments and a lot of people becoming wealthy from working with biometrics, but the field of biometrics did not evolve or progress [not] for a period of around ten years. Why? Well, because the people who got the money didn't do anything to advance the research on biometrics – they implemented whatever – or they simply didn't implement anything at all.

While Richard's criticisms should, of course, be understood in relation to his specific position as the head of a southern European biometric research laboratory, Louise Amoore also addresses the problematic relationship between economics, security and tech-proliferation post-9/11 in her book *The Politics of Possibility* (see also Chapter 3). In the book, Amoore explores the practices of authorization that have made it possible for private consultancies, risk management, software and biometric engineering to gain traction as expert knowledge. Amoore finds that after 9/11, a particular connection between economics and security directed at governing potential, emergent and uncertain futures was intensified. Drawing on Foucault and the American economist Frank Knight, she argues that the economy 'is always and inescapably concerned with the unfolding of future possibilities' (Amoore 2013: 5/10). Indeed, she considers that the whole idea of profit entails imagining potential future markets and avenues for income, as such being modelled on the unknowability of the future (ibid.). Providing several examples of how audit consultants and businesses working with the algorithmic modelling of customer behaviour changed their focus to the issue of security following 9/11, she argues that alliances between a commercial economy that embraces risk as potential and opportunity and the anticipatory decisions and pre-emptive approach to risk in state security have flourished. She explains that this alliance in terms of the absence of what she calls a 'definable central agency', an absence, which paved the way for the use of what was previously consumer tech in the context of security (Amoore 2013: 11). It was, Amoore argues, 'the exceptional circumstances declared after 9/11 that opened up the possibility for experts in the

unease of global markets to become the experts in the unease of state security'
(Amoore 2013: 21). Contrary to how it is often portrayed in the media, however,
it is not biometric technologies as such that are new (see also historical section in
the general introduction) but, rather, their authorization in new contexts (ibid.).
This post-9/11 situation has led to a proliferation of risk management consultancies, biometric companies and demands for biometric engineering skills in security, including border security and the management of mobility and migrants, who
are also interpreted as a security problem (Bigo 2002: 63).

Biometric researcher enskilment and mapping the pathways to resources

Although the contemporary use of biometrics may be entwined with a new type
of security thinking that casts migrants as potential security threats, this is quite
remote from the ways in which researchers in the labs I engaged with think about
their work and their daily practices. As mentioned in the site introduction, more
than anything the researchers think of their work as something that will contribute
to a 'seamless' future world, meaning a world in which everyday activities are
made easier and smoother through the use of biometric technologies. From their
participation in different events in the biometric landscape, however, the researchers know that security thinking is prevalent and often inscribed in project-funding
mandates.

The researchers in the two labs where I did my fieldwork attended a cycle of
events in the form of conferences and meetings throughout the year. Participation
depended on the relevance of the content of these events to current or possible
future research and the opportunities to network and promote their research results
and interests. Some events were generally prioritized and attended regularly by
one or two researchers. Among them were the meetings of the standardization
committees and the conferences and gatherings of the two large biometric associations. The often excessive prices charged for attending seminars, workshops,
meetings and conferences particularly on borders, security and biometrics, hosted
by private events companies, made it obvious that not all events were aimed
broadly at participants with an interested in biometric technologies. Conferences
were often held in fancy and expensive hotels, apparently being 'designed to be
exclusive and limit admission to organizations with sufficient financial capital'
(Baird 2017: 4). Therefore, not only biometric researchers but also myself as the
ethnographer had to consider carefully which events to participate in (cf. Baird
2017). Events were designed as places where industry representatives interact
with potential users and buyers, including special invite-only sessions and closed
workshops for particular actors. As such they also constituted sites where 'knowledge of practice' and processes of enskilment could take place under conditions
set by 'access categories' – that is, predefined categories of actors and institutions
that would be granted access to the events (Baird 2017: 4). In this way, knowledge in these settings became a scarce resource that was commodified and shared
among the select few and their friends.

For the younger researchers, such events were, in fact, places of 'enskilment' (Palsson 1994; Ingold 2000; Grasseni 2010). As Julia, who also had experience of working in the biometric industry, expressed it,

> One of my most important tasks is to teach the young researchers that it is not enough to make a computer code, a program or a sensor, just 'for the hell of it' – you need to keep an eye on potential results and applications.

She explained that university researchers need to learn to think in terms of collaboration and funding applications in order to 'make it' and continued':

> So this is why I like to take the younger researchers with me to different meetings. They may not be able to contribute, but they listen. They hear what is required, and they get to understand the system, the negotiations. This is really important!

She continued:

> In the beginning Richard took me to all the meetings on biometrics, thousands of meetings! On economy, on project requisites etc. I had to keep quiet because I really had no clue what was going on, but now I feel comfortable in these settings. I know what they are about, how to manage the stakes, how to socialize and understand the people present. This is important!

In these settings, the younger researchers had to listen and learn, becoming accustomed to the biometric landscape and acquire the skills to map the variety of important players, networks and forums for future use.

One of the sites where such processes of enskilment take place, and probably also financially the most accessible one, is the yearly conference on biometrics, organized by the Biometric Association on Seamlessness and Security (BASS). It is to this conference that I shall briefly turn next.

> *The large lecture hall at the biometric conference is attended by two hundred mostly biometric researchers, including several from the labs where I did my fieldwork. The conference at large gathers researchers, policymakers, practice-oriented research project staff, privacy experts, members of different EU organizations and representatives from the US Department of Homeland Security, an institution which, as Baird also mentions, is widely represented at European biometric and border conferences (cf. Baird 2016).*
>
> *The first speaker of the day is Aaron, a representative of the EU agency eu-LISA, which manages large-scale IT infrastructure, including biometrics. The agency defines biometrics as 'essential instruments in the implementation of the asylum, border management and migration policies of the EU' (eu-LIISA homepage).[1]*

> *During his talk, Aaron underlines the challenges facing what is known as the 'smart border system', including the biometric border installations placed at EU border control points. He proceeds to tell the gathering of mainly academic researchers how the agency draws not only on biometric companies' research and development but also on more academic research. 'And therefore', he explains, 'The future of EU smart borders is dependent on your work! We need quality algorithms and new ideas, for example, as to how we make the systems interoperable' (i.e. work together).*

This direct appeal from a representative of eu-LISA is an example of how the political agendas of EU member states, the biometric industry and biometric research are interdependent and feed into and off one another. Scientific lab work and 'the outside' are, thus related in a broader 'community of interests', which entails different types of stakes, where competition over market shares, contacts and funding is, at times, fierce and trading secrets abound but where researchers, industry and policy-makers are also deeply entangled in reciprocal yet unstable, contingent and sometimes conflictual relations. This became clear to me several times during my fieldwork, for example, when I was about to move from one laboratory to another.

Pathway 1. The wild and invasive species

In 2011 two biometric associations, the International Organization of Biometrics (IOB) and the BASS, formally initiated their work in Europe. Whereas the IOB came out of a non-European institution that expanded and set up an office in a major European city and later expanded to the US, BASS came out of a thematic network, including the European Biometrics Forum, which was funded by the European Commission until 2011. This association still has close ties with the EU system, meaning, among other things, that the association has influence on the EU biometrics agenda. For example, the association was asked to provide inputs to themes and content for the largest EU-funded research and innovation program, Horizon 2020. Furthermore, BASS has maintained a strong research orientation, even though EU policy-makers, data-privacy representatives and national security and industry representatives are present at the conferences that the Association organizes. The IOB, by contrast, seems to maintain stronger links with government officials and industry. These two biometrics associations are nonetheless important players in the field and host a number of conferences and events on biometrics across the EU each year. The rivalry between them, however, is rather intense, as they both aim to recruit as many prominent and powerful people as possible and they compete for human and economic resources, as well as over which of them has legitimate knowledge of and experience with biometric technologies and their current and potential uses. This rivalry also has implications for the work in the labs and the relationship between labs, as we shall see in the following.

In the ID-tech lab, I begin to sense an unease about me going to spend time with researchers in another laboratory in a different EU country. Richard frowns ever so slightly the first time I mention the people in the other laboratory, and several of the other researchers comment on my upcoming trip. Tara, one of the researchers, says jokingly: 'So I heard that you are going over to "the dark side"'. When I ask what this is about, I am told that it is meant as a joke but that it also has to do with the fact that the other laboratory is very good at making itself visible and getting good contracts, in spite of not always being the leading lab in the areas which it claims as its expertise. Furthermore, the other lab has been accused of preying on other researchers' ideas and successfully making them its own. I receive several similar comments over the following days, and I remember that a couple of months earlier, at a border security conference in Italy far away from the labs, I had a conversation with Mark, from a large American consultancy. Mark mentions that he is a member of both of these biometrics associations in Europe and that he will go to the next conference of the IOB.[2] 'But don't tell the people that you know from BASS', he says with a smile; 'they would not like it'. I think little of this conversation until the experiences in the laboratory and the realization that the researchers at the two labs where I conduct my fieldwork have chosen to affiliate with rival associations.

As my fieldwork progresses, I hear different stories about the rivalry between BASS and the IOB in – and outside – the labs. While talking to Richard about the field of biometrics in Europe, he also mentions the relationship between the biometric associations. Knowing my interest in migration, Richard compares the rhetoric between the associations to debates on migration, within which he positions the associations as either 'migrant invasive species' or 'racist and discriminating host countries'. It got to the point, he says, where the prevailing attitude was one of 'You are either with them or you are with me, and if you are with them you are against me'. Richard is a member of both associations, but he felt that people had started acting strangely around him because of this double affiliation.

These fieldwork experiences and stories of tensions could be understood as based on simple rivalry and jealousy between two research institutions. However, the story is also part of a larger process involving the right to define what constitutes legitimate knowledge of biometric technologies and who embodies this knowledge, as well as the construction of not only national borders but also of borders between those who legitimately constitute the European 'biometric community' and those who do not. The emotions invested in the rivalry between the two associations do not only have financial and academic implications for the researchers; they also seem to involve potential inclusions and exclusions in relation to particular relational settings.

In a complex, conglomerate, but also rather small field, of expertise and research, the competition for attention, interest, legitimacy and, ultimately, funding requires that the researchers continually work on promoting their scientific achievements and show how these achievements might become useful in providing new aspects of border security. This is also the way to ensure sufficient funding for continuous basic research, as mentioned in Chapter 1. In order to expand networks, the laboratory researchers sometimes attend events hosted by both IOB and sometimes by BASS, and it becomes a strategic choice where to attend when, and when to disconnect and 'cut the network'. Such strategic choices are ultimately reflected in the project portfolios and daily work in the labs. As Richard tells me, the project on vein biometrics which the lab initiated with an Asian government was established through the networks that he had built through one of the important European biometric forums, the IT security association TeleTrusT, a large, non-profit interdisciplinary competence network for IT security experts founded and based in Germany but active all over Europe. This is also an association that harbours close ties with BASS by virtue of sharing several of the same prominent members.

Pathway 2. Negotiating the value of biometrics

A meeting room in a luxury hotel in a European capital in 2017. The strong smell of aftershave is striking. An almost all-male audience, immaculately dressed in dark, with well-polished shoes and shiny slicked-back hair, fills the first row of a biometric workshop being held at a border security conference. The workshop aims to introduce biometric technologies, explain their business drivers and demonstrate the purpose and use of biometrics to prospective users and thereby also to potential buyers. Among the participants are the Greek and Moroccan border police, European navy and EUROPOL officials and representatives from the immigration services of an African country. The event is also attended by biometric vendors, as well as the US agency OBIM (Office of Biometric Identity Management') under Homeland Security, which already makes intensive use of biometrics. The presentations are being given by Antony, a biometric expert and member of BASS, Hermann, a German researcher from the ERC (European Research Council), and Mark, the representative of a US-based biometric consultancy mentioned earlier. Antony, the biometrics specialist, tells us about the business drivers: security, traveller convenience and economic efficiency and how different stakeholders have different views on the right balance between these three drivers. The ways in which biometric installations work are therefore always the outcome of intricate discussions among, for example, airlines, airports, border guards and police harbouring different interests (see also Part II of the present book). During this introduction to biometrics, a man sitting at the back of the room, who obviously knows the speakers, is continuously challenging Antony on his views of the limitations and potential contingencies of biometric installations with comments such as 'the industry is already remedying this' and 'this is not a problem of biometrics, but of border guards'. Both Antony and the other speakers know him and address him by his name, Philip. He turns out to be a representative from a biometric vendor named 'Stargates Inc.', which is quickly spreading its activities in Europe and beyond, particularly in the area of border control.

During a break, Philip comments that he finds the atmosphere too critical of the technologies, and after a presentation from Herman, he adds, 'It's typical; the Germans are always so critical!' In contrast, Herman, the German ERC representative, in a discussion during one of the breaks complains to me about governments that suffer from technophilia and the need for quick fixes. He adds:

There is a huge problem with vendors who assemble off-the-shelf products that do not communicate well. Consequently they never become well-functioning systems. Vendors tend to oversell their products in performance assessments, and they know it! 'Give me a performance claim, and I will give you a dataset that proves it!' And with the extra anti-spoofing applications the systems are becoming ever more complex and more difficult to operationalize, but border guards should not have to be biometric experts! And not everything can be solved through technologies. Still, there is a trend among politicians towards believing in technologies instead of training the border guards whose experience and knowledge should be the focus instead. The *real* verification should be done by the border guards. Biometric technologies *will* make the borders insecure if applied uncritically.

Herman depicts the biometric landscape as being under the heavy influence of technophiles and their belief in, and wish for, technologies as security 'quick fixes', as well as an industry that is mostly interested in promoting its goods. Herman himself is positioned firmly in an EU research department and has dealt with migration and biometric identification for many years. Conversely, Philip, the head of marketing and sales in Stargates Inc. who came to the field of biometrics from the broader security industry four years ago, sees a field plagued by unfounded fears and concerns that prevent the technologies from developing satisfactorily, and he paints the industry in a bleak light accordingly. What becomes clear here are two of the main positions and controversies inherent in the field within which the laboratory researchers need to learn how to navigate. The latter are aware of the limits of biometric technologies and the potential pitfalls of 'technophilia', as Piero from the lab argued:

> No matter how safe we believe a technology to be, it should never be more than a tool – the tool cannot make the final decisions in important matters, and there *always* has to be a human involved. Biometrics should be used the same way that we use a knife. A knife is capable of cutting meat for food, but if a person is not using a knife – then it will not do anything. That's a thing we have to be aware of when we work with biometrics.

The work of the lab researchers, however, is obviously not helped by promoting 'technophobia' or by excessive caution. Thus, taking up an intermediate position by separating the technical from the political, economic and ethical domains becomes one way of handling this dilemma for the researchers. Piero goes on: 'Biometric Technology is a tool, and it should always be a tool – it should never be in charge! Whether this tool is good or bad depends on how, why and where it is implemented'.

One forum where such actual use of biometrics and the contexts of implementation are discussed among multiple stakeholders is in the standardization committees.

Pathway 3: Negotiating standards, enabling technologies

At the first encounter with it, standardization work comes across to the novice researchers more like a jungle of different institutions, technical committees, sub-committees and multiple working groups and forums than an ordered site. On their path through these forums, the researchers must also become acquainted with a particular type of language characterized by a standard-technical jargon, as well as references to previous and international work on standards, consumer rights, legal issues, intellectual property rights and so on. This is also a site where controversies unfold and where the assembled and differently positioned actors are trying to reach consensual agreements and thus apply order, stringency and coherence and, in other ways, 'round up' biometric concepts and practices that are 'running wild'. One of three collaborating bodies of standardization recognized by the EU is the European Committee for Standardization (Comité Européen de Normalisation/CEN), founded in 1961 with the aim of developing EU standards in different sectors, in CEN's own description, in order to ensure common product standards, facilitate a common market and place Europe in the global economy.[3] Since 2010, European work on standards includes a working group on biometrics that aims at developing consensual agreements about standards on a number of issues related to biometrics.[4] These include the content, vocabulary, use and meaning of biometric technologies and data formats applied in biometric systems, as well as consumer rights and data privacy.

As we shall see in the following, and as noted also by Linda Hogle (2009) in her article on tissue engineering, the work on standards highlights the way in which 'political–industrial [and here I would add scientific] assemblages participate in socially negotiated forms of objectivity and are inseparable from the way new technologies take shape' (Hogle 2009: 717).[5] A big challenge in this type of work is to develop standards that meet all the different and sometimes contradictory interests and positions of the participants who represent the different sectors. The participants in the working group on biometrics consist of members of national standardization boards, representatives of biometric companies, researchers from university departments, biometric experts (people with many years' experience in the field of biometrics and standardization), representatives of the EU border agency (Frontex) and representatives of the European Association for the Co-ordination of Consumer Representation in Standardisation (ANEC). The outcome of this work on standardization is mostly embodied in official documents describing standard procedures to be followed, recommendations and best practices in relation to several aspects of biometrics, including recommendations for finger-print hardware and software, user enrolment and information in biometric border gates, ways to increase efficiency in border security and ways of ensuring the systems against possible spoofing or presentation attacks. According to Hogle, such consensus-based documents, which are produced by intense and 'socially negotiated forms of objectivity', seem to acquire fact-like status once they are recognized as international standards (Hogle 2009: 717). They are, however, at least in the context of biometrics, amenable to revisions over time.

Several researchers from the laboratory participate in standardization work at different levels, from CEN to the International Organization for Standardization (ISO), as well as the joint technical committees with participants from different standardization bodies.

Raphael, from the ID-tech lab, who often participates in these meetings, told me that he initially thought they were rather tedious and long and held in an extraneous language. However, he now thinks differently about them he tells me:

> After about two years I finally understood the idea with the meetings and got to learn the vocabulary. In the standardization meetings you get to meet and know a great variety of actors that play in the biometric field. You make yourself and your research visible, *and* you learn about the field and what is going on there in terms of development and projects. Furthermore, there is a tight fit between the granting of EU funding and standards. The EU develops standards *from* EU projects, but also recommends that the projects themselves conform to the already available standards. So, if you want to take care of your interests, this is a very important forum to participate in.

In a conversation on standards with Peter, a professor of biometrics, he remarked that the standards are also 'good for society', since they aim to ensure that biometrics live up to certain regulations, are safe and reliable, and that consumers' interests are being heard. Furthermore, he added, the work on interoperability in the committees helps protect against the practice of designing systems that lock customers to a particular provider's soft-/hardware (vendor locks).

Hearing about the work on standards in the labs and at conferences, I decided to join Elsa, one of the lab researchers, at a standardization meeting on biometrics held in Germany. The following field notes come out of this meeting and show the intense social negotiations and positioning that take place in this context:

> *The participants in this European standardization working group meeting represent a range of professional profiles: a biometric expert, two university-affiliated researcher groups, four vendor/industry representatives from different EU countries, a consumer representative and representatives from national standardization bodies. Today in the meeting room members from Norway, Spain, Finland, France, Germany and the UK are present, and it is clear that, as at the conference, most of the participants know each other rather well.*
>
> *We go through a document on the vocabulary of 'attacks on biometrics', which has been circulated between the different national delegates for online comments. As we go through the sentences in each paragraph minutely in order to reach a consensus on the wording, it becomes clear why this is such an important forum. The different types of words chosen allow more or less freedom of interpretation and contain inherent decisions concerning questions such as whether or not 'a newly grown beard' should be written into the document as one of the face modifications that biometric technologies should*

*be attuned to. Another question addressed, for example, how and when trav-
ellers going through biometric gates should be informed about the use of
their personal data. As an opinionated person I cannot help but throw in
a comment here and there, although, as a guest, I have no formal power in
the procedures. Seeing slight changes to the vocabulary following my comments, however, gives me a surprising sense of empowerment. Am I actually
changing a sentence in what might at some point become an international
standard?*

*During the meeting a controversy breaks out between a (German) con-
sumer and privacy expert and a biometric expert. The controversy concerns
whether or not a biometric border system should be understood as a system
that, alongside its registration and/or identification of border crossers' biometric features, also assesses whether those crossing a biometric border are
trying to 'attack the system' or not.*

*The privacy expert argues: 'The users need to be told exactly what their
data is being used for'. The biometric expert and researcher answers: 'But
these two things go together – checking for attacks and checking the biometric features are part of the same process'.*

*The privacy expert raises her voice: 'No they are not! It's not the same if
the system checks whether the traveller is an attacker or just verifies their
biometrics! And the user needs to be told what their data are used for'.*

*The chairman, a representative of a major biometric company, takes the
floor: 'So when do the users need to be informed?'*

*Privacy expert: 'Well, they need to be informed just before they enter the
biometric gate when you use their personal data like that – and biometric
data per se is personal data'.*

*The chairman continues: Hmm, OK . . . Well, but not a lot of systems are
compliant with [i.e. live up to] this'.*

*Privacy expert: 'Well, but you need to write the obligation to inform about
data usage into this document. And this needs to be a requirement in order to
install a system'.*

Everyone in the room becomes silent.

*The privacy expert goes on: 'It is very clear that the users – the travellers –
need to know what their data is used for'.*

*Biometric expert: 'I do agree to that. People should know what their data
is used for'.*

Chairman: 'Well, I think it is implied, but OK, we will make it more explicit'.

Here, in a matter of a few minutes, the standardization committee has 'in principle' refashioned a small aspect of the border arrangements at EU airports, many
of which already have biometric gates installed. In practice, from this point on,
materials designed for airport biometric gates should inform travellers of the exact
use of their biometric data, a space should be made available in front of all biometric gates and instructions must be issued to border guards or police officers so
that they can answer potential data-handling questions from travellers. This is not
a quick process, however. The negotiations of standards take time, and most of the

biometric standards, like in this case, are worked out and applied 'ex post facto'. The example, however, makes it clear why not only biometric researchers but also biometric vendors consider it important to be members of the standardization committees. Thévenot (2009) argues that, apart from their aims in providing regulation and objective categories, participation in such forums constitute an 'investment in form' (Thévenot 2009: 794). The return on this investment depends, he argues further, on the 'temporal and spatial validity of the form, and the solidity of the equipment involved' (ibid.). In other words, it is a matter of how long the standard that governs; for example, a particular biometric fingerprint technology is at work in the community of users and how extensive this community is. This, Thévenot argues, is why participation in such meetings, which are preoccupied with the elaboration of 'form' – here in the shape of biometric standards – provide a way of preventing standards from 'becoming external to one's own concerns' (Thévenot 2009: 794). In the standardization meetings on biometrics, for example, the vendors are able to negotiate and thereby try to ensure that the products they *have already* developed will not suddenly be asked to live up to new standards and consequently be harder or impossible to sell. Researchers, in their turn, often provide the algorithms and programs for commercial or state-sanctioned products, as we heard in the initial empirical quote by Aaron from eu-LISA, and the standards may therefore also have direct implications for their work. Furthermore, although the researchers broadly confess that working in these committees is sometimes tedious and slightly boring, they also highlight the need to stay attuned to the discussions in the field, as well as the potential for collaborations and alliances that come out of this work. The standardization committee, then, is a forum where decisions, however minute they may seem, have consequences for the ways in which technologies can be and are implemented at, for example, airports, and how they may be modified and worked into existence in both research and commercial vendor labs. It also constitutes a forum where researchers, vendors, privacy experts and national representatives attempt to promote their own agendas by establishing different types of relationship: liaising, developing joint projects and obtaining news about particular topics that are of interest and that could help define important themes for applications for further research funding.

Vendor fairs, such as the 'passenger terminal expo', a yearly conference focusing on different aspects of airport security and convenience, constitute another type of event where such types of knowledge are produced. With more than 7,000 participants from more than 100 countries, according to the organizers, this is an important site for the promotion and sale of biometric technologies. Here, biometric companies' research and development staff meet, for example, airline representatives, security and other airport employees and decision-makers who have an interest in products that may enhance security and/or convenience.

Pathway 4. Unveiling and concealing

> *The expo hall with its crude artificial lighting is immense. Multiple desks and booths set up by particular companies selling their products are visible from the large entrance. There is an ongoing murmur and hum of voices from the audiences*

*that are circulating and the soft-speaker voice-overs coming from video presen-
tations of particular products. Different types of biometric border gates are displayed, they entail different mediations between speed and security and require
more or less bodily contact with the inbuilt sensors as they flash in different
colours and capture faces, iris, fingerprints – even veins. Big screens show live
recordings of airport maps with people on the move as red, green or yellow dots.
Even the robot Cahiba, which is meant to assist travellers, is on the move in the
large hall, closely monitored by its makers and vendors and a crowd of curious
and amused onlookers, while the staff members at the exhibited biometric gate
across the corridor snort condescendingly about the intense interest in this 'useless toy'. After all, who needs a robot in the airport if everything is signposted
through large interactive and personalized screens?*

Welcome to the first day of the airport expo. This is one of the places where biometric software and hardware is not only sold but also negotiated and fiercely protected.
I am visiting the company Stargates Inc. that Philip, whom we heard about previously, works for. Stargates has an exhibition at this fair. They have just won a big
contract for a border gate system with a large EU airport and are celebrating. Their
systems are assembled from different types of programs, sensors, camera lenses,
facial recognition software, glass, metals and other types of hardware. Some of
these components they develop themselves; others they buy from other companies
or develop in collaboration with or on commission from biometric labs. They are
not alone, however: other companies selling similar products are present at the fair.
I think little of this, and after the first day's conversations with the staff of Stargates I
move freely around the expo hall, posing curious questions to the staff at the various
stalls. I return to the Stargates stall regularly during the three days, talk to staff about
their work and the biometric installation they have on display. I hear them speak to
prospective customers and we small talk about biometrics, about how they believe
that their systems are better etc. I do this mostly in an attempt to find out whether I
might be able to do part of my fieldwork in the company and how.

Toward the end of the third day, when I return to the stall, I sense that something is going on. The Stargates staff members are suddenly no longer so talkative
and friendly – they seem to avoid eye contact, and when at the end of the expo
I come back to say good-bye, I find them assembled in a circle talking in low
voices. They stop their conversation abruptly as I approach. I ask if there is a
problem, and after initial murmuring, a male representative, whom I have not seen
the previous days, steps out of the circle:

> Yes! I have worked in security for many years, and I want to know why you
> are here, who you have talked to, what you will use it for . . . and how we can
> know that you are not sharing this information with our competitors?

I am thoroughly surprised and try to reiterate my motives for being at the expo
as best I can. The incident, however, rather than being just embarrassing, serves
to highlight the extreme competitiveness of the biometric landscape. On the one
hand, events such as the airport expo are all about making the companies and their

products highly visible and sellable and answering intricate and detailed questions from potential customers. On the other hand, they are also about making sure that information is not given away that can be used to enhance other similar products. This makes such events a careful dance between *unveiling* and *concealing* product info and new inventions. There is, however, a limit to the secrecy that companies and research labs can actually maintain. EU standards, which, as we have seen, come out of the compromises reached at standardization meetings, increasingly demand that different biometric systems and interfaces are able to work together, share data (i.e. are 'interoperable') and avoid 'vendor locks'. As a result, too many trading secrets and proprietor algorithms potentially become an impediment to the acquisition of EU funding.[6]

Back in the lab

Back in the lab, Richard is keeping himself busy, always on the lookout for new funding, new contacts and new projects, which is ultimately what allows his biometric laboratory to exist. He sits on several of the boards of biometric associations and attends conferences and meetings with current and prospective project collaborators, as well as with companies that might want their biometric terminals evaluated or would like to apply biometric access control somewhere in their work processes. He knows the biometric landscape like the back of his hand. Apart from these tasks, Richard also has to take care of his teaching obligations, among them teaching the police about biometrics – how they work, how they are useful to police and border and security tasks and so forth. It is, in principle, Richard who defines which biometric technology the lab should engage with. However, the two other senior members of the research team mentioned earlier – Julia, who worked for one of the largest biometric companies in the EU and participates in several EU projects, and Rafael, who regularly attends biometric conferences and focuses on user perspectives – influence these decisions equipped with their knowledge of and links to the biometric industry and current funding landscapes. By bringing the younger researchers along to the different events and activities that take place in 'the biometric community', they also ensure their younger colleagues' ability to navigate in this ramified landscape. Apart from the desires of researchers to push the frontiers of science, their mundane 'small decisions' and the practical circumstances of the lab depicted in Chapter 1, the nature of the research that is set in motion in the lab and the biometric technologies that crafted there also depend on this collaborative knowledge and the skilful navigation of the biometric social landscape. It requires the mobilization of interests in lab work, the proficient mapping and reading of political interests and anxieties, as well as the ability to tap into policy-driven dreams of a safe and seamless future and to imagine the potential sites and situations in which biometric technologies might be applied.

Biometric labs in the border world

Part I of this book has shown how biometric technologies are fashioned at the intersection between laboratory research interests and a wide range of circumstances and social and political dynamics. By using cartography as an analogy and conceptualizing the researchers as 'body cartographers', I have argued that, like curious cartographic explorers of unknown territories, they develop maps and experiment with mapping strategies, which, however, metrify, abstract and transform lively body territories into maps that enact bodies as particularly bounded, stable and static entities. In a migration context, these maps are used to regulate movement by virtue of locating the border on (or inside) the body. I have argued that biometric researchers strive to be at the forefront of science and that imaginaries of what may seem like distant and, at times, rather extraordinary technologies and science-fiction scenarios are explicitly aspired to and worked upon in the labs. Some of these technologies, as well as the future visions that are attached to them, are also connected to political agendas and dreams of a secure and efficient future. Finally, Part I has shown how multiple connections and disconnections are mobilized in laboratory work. This occurs in the daily work and 'small everyday decisions' on how to disconnect body attributes from the bodies they are normally part of and turn them into mobile digital maps that can be connected in biometric systems. Furthermore, different types of connection and disconnection are made whenever researchers move from the lab further into the biometric landscape. Here, connections are mobilized and disconnections effected in order to establish what are the legitimate sources of biometric knowledge production and the technological applications that may support claims to project funding, thus ultimately ensuring the continued existence of the individual research labs.

Notes

1 The European Agency for the Operational Management of large-scale IT Systems [. . .] (eu-LISA) is set up to provide a solution to manage large-scale IT systems. The management board consists of representatives from the EU member states and the European Commission, Euro-just and Europol. The current Danish representative in eu-LISA is also the head of the Danish immigration service (Udlændingestyrelsen).
2 Mark is also on the board of their 'vulnerability group', which, rather paradoxically, does not work with migrants' or citizens' vulnerability to biometric registration, but instead with the vulnerability of biometric systems and the avoidance of potential attacks against biometric systems themselves.
3 See also www.cen.eu.
4 The International Standardization Committee (ISO) also has a subcommittee on biometrics that has existed since 2002.
5 The work on standards also highlights international rivalries in biometrics. As I was told during my fieldwork, one of the most important tasks of the European forum for standardization is to foment a strong European voice on biometrics, which can provide an alternative to what is perceived as a rather US-dominated field.
6 'Interoperability' is also controversial, however. Arguments against interoperability are sometimes rooted in concerns with the protection of user data. By preventing interoperability, data sharing is potentially more difficult, and so are the possible infringements of data privacy.

Epilogue

When the French police officer Alphonse Bertillon developed his anthropometric system of body metrification around the 1890s, known as 'Bertillonage', little did he know that his system would be challenged and later largely abandoned as a result of one of those 'wild' things of nature – a pair of identical twins being taken into criminal custody years apart. The twins had the same body metrics and were therefore impossible to tell apart by using Bertillonage. It took fingerprints to differentiate them. Bertillon was subsequently forced to make a space in his already very elaborate body metric formula for a fingerprint, but even so, the fingerprint would gradually supplant the very time-consuming and less accurate approach of Bertillonage itself.[1]

Fast-forward to a rainy August morning in Twinsburg, Ohio, in 2009. It is the first day of the Twins Days Festival. Since 1976 biological twins, triplets and quadruplets from all over the world have attended this weekend festival. During the festival identical twins are supposed to dress alike, wear the same hairstyle – in sum, look as identical as possible. This year the festival is attended by at least two participants who are not twins. Two biometric researchers from the University of Notre Dame in Indiana funded by the Federal Bureau of Investigation have set up a research tent in order to answer the question, 'Can the best state-of-the-art facial recognition distinguish between two genetically identical people who are trying to be indistinguishable?' (Strickland 2011: 1). Facial recognition does not perform so well. 'Identical twins represent a real torture test for biometrics', one of the researchers tells the reporter Eliza Strickland (ibid). When the photos of the twins were taken under ideal conditions, with neutral expressions and under studio lights, the facial recognition software did reasonably well. The programs were not completely confused by the similarities. However, in order to generate a more reliable and higher recognition score, the researchers felt that the system would have to be fed higher-resolution images and be able to capture minute details, such as moles, rather than simply the relationship between facial features, such as eyes, nose and mouth. When the facial recognition was tested in the wild, outdoors, with faces showing different expressions, the system worked very badly (Strickland 2011).

Fast-forward to the 35th Chaos Communication Congress in Leipzig in 2018. In front of the assembled journalists, hacktivists and other interested participants, the hackers and computer scientists Jan Krissler also known under his alias 'star-bug', and his colleague Julian Albrecht managed to spoof state-of-the-art biometric vein scanners made by Fujitsu and Hitachi by using a modified camera and an artificial hand made from yellow beeswax. These technologies were promoted as among the safest products on the market. While the attack was no quick endeavour and required time and expert knowledge, it was inexpensive and successful.[2] The work by hackers such as starbug, and especially by unknown and unpredictable hackers and spoofers, constitutes another form of 'wild' that the biometric researchers continually have to integrate into their research efforts.

'The wild'

Part I of this book has taken us on a tour of the biometric community, departing from but not limited to two biometric labs. However, the technologies produced through multiple engagements described also have to be implemented in what figures as 'the wild' when seen from the labs, for instance, at land, sea and air borders, in ATMs, in cars and mobile phones. As I have shown in the previous two chapters, there is no clear boundary between the lab and the wild when it comes to how biometrics are engendered. Instead, researcher aspirations, political and economic interests and conjunctures travel as impulses along pathways that connect the labs to the biometric community. 'The wild', however, remains an important general trope for the world outside the controlled environment, as well as a way of addressing the unforeseeable circumstances and occurrences that may have implications for the working of the biometric technologies once they are 'let loose'. The unforeseeable is therefore also something that the researchers in the labs spend their time trying to 'rein in' through, for example, the scientific literature and not least through their own imagination and experimentation.

The work in the labs and the kinds of technologies developed there are not only an outcome of biometric researcher's research agendas or of the political and economic structures and cries for security. The development of these technologies is also an outcome of the 'wild' and uncontrollable phenomena, the people, environments and practices that exist outside the laboratory that require the laboratory researchers' attention. It is exactly such 'wild' sites to which we turn in Parts II, III and IV of this book.

Notes

1 Apart from other body metrics, such as the cranium, profile, nose, height and so on, Bertillon worked on measurements of the ears. Interestingly, ear biometrics is a modality that is currently being developed further, although it is not just based on the metrics and shape of the ear but also on the individual reflection of sound waves that rebound off the tympanic membrane when a sound wave is sent into the ear cavity.
2 Already in 2015, however, researchers at a European research laboratory had written an academic article on the possibilities of spoofing palm-vein biometrics (Tome and Marcel 2015).

Part II
On the border

Perle Møhl

Introducing the site

Every border is characterized by a particular constellation of disjunction and contact, cross-border agreements, contestations and friction, as well as technological infrastructure, with often a deep historical and geopolitical anchorage in the landscape. To those who pass them unhindered, borders might be an example of Marc Augé's 'non-places' of pacifying transition, solitariness and detachment (Augé 2008). To others, as Michel de Certeau suggested (1984), borders can be significant sites of intense activity, interaction and human encounters, whether they constitute an obstruction of one's route that needs to be negotiated or a workplace, a site of routine choice-making where small decisions are made around the clock. Technologies play an important part in decision-making, and technological expertise is performed by those who are there to protect the border, as well as by those who study and seek to circumvent it by developing technologies and skills of their own.

As we describe the border world in the Introduction, national borders can be externalized to the farthest corners of the world, internalized in bodies and delocalized as mobile data sets on servers. The border in Part II, however, is an actual physical border, patrolled and surveyed by border guards and technologies. Where borders and border work in Parts I, III and IV concern abstract, delocalized or imagined borders – for example, imagining future procedures for filtering bodies, fingerprints that curtail one's movement, and rules defining the 'proper' family – Part II is concerned with visible border installations and 'identifiable and locatable actors' (Pallister-Wilkins 2017: 64). And whether borders are physical places or ephemeral technologically mediated instantiations, they are in all cases produced through practice and *take place* – in Part II, very tangibly.

In analysing the integrated *border work* of border guards, migrants and technologies at physical borders, Part II seeks to elucidate the daily routines of human-technological decision-making processes, the bases on which they take place, and how they contribute to characterizing the broader border world assemblage. Part II is especially concerned with the functioning, proficiencies and limitations of particular types of biometric technologies that are deployed on the border in combination with human sensory work and interpretation. It does so by looking at two types of European border – airports and land borders – involving different kinds of travellers, guards and encounters, different types of choice-making and different types of technologies that make or help make those choices. A particular focus is on the interaction between border guards and technologies, the use of human senses and interpretive work and how human senses and technologies play together and mutually format one another.

The settings

Biometric border technologies vary from very simple tools like height meters to complex body scan and facial recognition technologies. The goal of this part of the research project was therefore to study border control settings where as many as possible of these technologies were used in combination with the border guards' own skills and sensory work. The goal was to compare how they worked in practice, how their human–technological interactions differed, and to look in detail at the series of infinitely small decisions, both human and technological, through which the border assemblage was instantiated. As a visual anthropologist interested in the senses, in interpretation and in the frameworks – technological, political, organizational, material – that organize the daily interpretation of signs and decision-making, I was especially attentive to the visual and sensory aspects of border work. In particular, I was interested in how the border guards were trained to use their senses and to see, both directly and via technological interfaces, in order to make decisions. I therefore started out by looking for sites like airports, where facial recognition technologies were being used. I also wanted to investigate the kind of sensory work involved in guarding huge material installations such as actual border fences in order to compare the different kinds of technologies, their relative efficiencies and the types of sensory work involved. In other words, the technologies and the particular border zones where they were being deployed guided my entry into the field.

With help from a colleague working for the national police, I was initially allowed to do a pilot project at Copenhagen Airport which laid the basis for a longer period of fieldwork (see the later discussion). The necessary police clearance, once obtained, then enabled me to move from one police force to the next, with a shorter visit to the Danish–German land border around Padborg (see the map in Figure 2.1) and then on to Gibraltar, where the joint Border & Coast Guard Agency received me for fieldwork in 2017.

Various types of biometric technologies were being employed at the airport and on the land border with Spain, along with classic radar and visual systems. From Gibraltar, I moved across the Strait of Gibraltar to Ceuta, a Spanish urban enclave on the Moroccan coast. Here, the border was manifest both as a huge double fence armed with a range of presence-detection technologies and in the form of the sea itself.[1] I was eventually allowed to work with the border guards of the Guardia Civil in Ceuta, but I also became aware of the highly developed (and fruitful) technological skills of the migrants who had crossed the border fence and were now being processed in Ceuta (see Chapter 4).

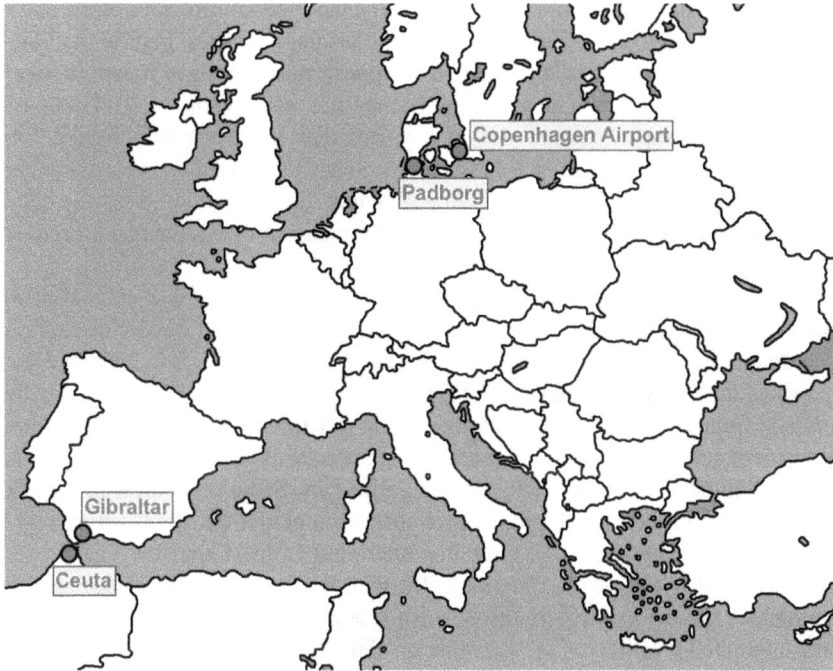

Figure II.1 Map of fieldwork sites
Source: By author.

The cases presented here are thus based on fieldwork among police officers in two international airports – Copenhagen Airport and Gibraltar International Airport – and with Guardia Civil border guards and migrants in Ceuta on the land border between Spain and Morocco. All are Schengen borders and function under the border control legislation of the EU coupled with national exceptions. But the practical settings and conditions under which they work are very different, the first two being airports inside a national territory receiving passengers arriving by air, the last being a physical border between two national territories with a fence or a sea providing the line of separation. The differences in these settings and in the technologies deployed provide insights into some of the variegated modalities of European deterrence practices, different ways of 'making sense' of an intrusion and different ways of circumventing the systems. In their different ways, they also display some of the incongruities of border maintenance and mobility regulation in and out of Europe, as well as the many forms of disconnection that rule both within and between these particular border worlds.

In Copenhagen Airport, border control became restricted to passengers arriving from or leaving for non-Schengen countries when Denmark joined the Schengen area in 2001; all other passengers could move freely within the Schengen area. Roughly 20,000 persons pass daily through the passport control zone at the entry to Pier C arriving from or leaving for non-Schengen countries, mainly the Middle East and Turkey, Asia, the US and European non-Schengen member states such as Ireland. As a heightened security measure, since October 2017 all passengers' documents are scanned, which means that no discrimination based on profiling will take place. Besides the regular Schengen border control, a number of 'random controls' take place on intra-Schengen flights arriving from so-called high-risk Southern European cities as a direct measure of systematized Frontex migration control.

In Gibraltar, the border is a highly symbolic and intense site of continued contestation between Spain and Britain, especially the land border separating Gibraltar from Spain that was de facto closed by General Franco in 1969 and not reopened until 1982 (Orsini et al. 2017). Gibraltar is an EU member but, like the UK and Ireland, is outside the Schengen area. Because of its prominent position at the entrance to the Mediterranean and its particular fiscal status, it has become a site of intense economic activity, categorized by Spain as contraband (Pack 2014). However, despite its prominence seen from across the strait, surprisingly few migrants have arrived in Gibraltar according to local sources. The border zones are highly equipped technologically, with facial recognition on both sides: Automated Border Control on the Spanish side and continued facial registration and automated profiling on the Gibraltarian side of the land border. At the airport, many passengers arrive with Schengen visas that do not give them entry to Gibraltar, but accommodating border officials occasionally organize unofficial transport to the Spanish Schengen entry just 50 metres from the airport instead of forcing them to fly back to their point of embarkation.

The history of Ceuta – a history of conquest and continued border-making – is inscribed in a complex Mediterranean history dating back to the Phoenician and later Roman Empires, the Visigoth attacks and the Umayyad Caliphate's conquest of Spain and subsequent Moorish rule over the region. Ceuta came under Portuguese rule in 1415 and Spanish rule in 1668 and has since been the site of both warfare and negotiations with Moroccan forces and governments (Saddiki 2012), becoming one of the European Union's external borders when Spain entered the EU in 1986 (Gold 2000; Pallister-Wilkins 2017). During this long history, the border itself has been made very manifest, by both a strong mural fortification of the city centre and, subsequently, the increasingly imposing fence along the border with Morocco. The fence is characterized by both its material and technological 'hard-wiring' (Andersson 2016) and by the fact that large groups of mainly sub-Saharan migrants regularly manage to cross it (see Chapter 4).

Getting there

Since the borders examined in Part II are actual tangible material settings, determining where to go is therefore not the problem. Acquiring permission to go there, however, is another matter altogether.

Indeed, being allowed to participate in and observe police work up close is not at all simple for a variety of reasons, as the relatively few ethnographic fieldwork-based studies of policing in Europe demonstrate (e.g. Andersson 2014; Fassin 2013; Feldman 2019; Hartmann et al. 2018; Holmberg 2003). As for border police using biometric technologies, to my knowledge, only a few anthropologists have conducted their analysis on the basis of ethnographic fieldwork with border police forces (e.g. Alpes 2015; Andersson 2014), as they work more often from reports, policy papers, interviews with officials and border security conferences (e.g. Kuster and Tsianos 2016; Maguire 2014, 2018; Schindel 2016).

In my case, I wrote to the chief of the Copenhagen Border Police and presented my interest in the relationship between humans and machines and in the interaction between their *different ways of seeing*. For the chief of the Copenhagen Border Police, this struck a chord, since his force had been grappling with such questions themselves. Management, police officers and unions were discussing the effect of automated vision and of automation in general for resource, wage and labour policies. The pressure on border work had been rising since 2015, when new EU and Danish regulations were introduced. Increased border control also meant that other areas of policing were being understaffed, so more civil border guards were being hired and were undergoing a relatively brief training program, thereby leading to the perceived devaluation of the work of the police officers on the border. Management, in my case, was probably interested in an empirical assessment of what was most efficient – human or machine vision – in order to make their case. The managers also, I was told later, considered it their duty to open their doors to researchers to show that they had nothing to hide. In all cases, they were sending a signal both to themselves and the outside world that this was an open institution, that they were taking border work seriously and that they were doing their work correctly.

At the airport: wearing a police badge, constituting a field

In almost all fieldwork settings, you cannot decide in advance what will constitute your empirical field. On the contrary, it is what you are allowed to participate in and the positions people attribute to you that define the contours of what will gradually become your field and what you will get to know something about (Møhl 2011). That is the predicament in all ethnographic fieldwork.

From this point of view, one thing very clearly defined my position in relation to travellers in both Copenhagen and Gibraltar airports: the fact that I was working inside a controlled zone and wearing a police badge. This obviously had important implications for my position and for what came to constitute my field. While I was expecting to acquire some insights into the experiences and the motivations that made people attempt to cross the border – especially those who did not cross it seamlessly but were held back for various reasons – it quickly became clear that I could not ask the questions I usually would as an anthropologist. Travellers identified me as a police officer, and I heard only the things that a police officer would: requests for information, mostly silence, and on one occasion a request for asylum – but never the more intimate discussions that an anthropologist can have with people about their goals and aspirations. And because I was observing the work of the border control agents and taking notes, I was ostensibly a senior officer in the police hierarchy overseeing and controlling the agents. This was an unusual and challenging position to be in and a predicament that I continually analysed and tried to learn from as a constituent of the border world itself. It clearly demarcated my field, what I could get to learn about it and what I could not. My field became a very tight time-space zone constituted by the border itself and its brief and direct encounters between police officers, travellers and technologies. In a certain sense, to the travellers, I *was* the border. That also meant that the ways in which travellers related to me and looked at me and in which I looked back at them became part of my empirical material.

In Ceuta: expertise and entanglements

My concern when moving into this very tense border setting was that my position with the police would not permit me to discuss border technologies with those who were trying to pass or circumvent the borders. This was, as already mentioned earlier, the case in Copenhagen and in Gibraltar, where I was considered part of the border apparatus. But to my surprise, this was not the case in Ceuta.

I had initially made a ministerial request to follow the Guardia Civil in its work. After a month of waiting, the request was granted, and I was allowed to accompany a senior officer on his patrols along the fence and at the entry point at Tarajal II, as well as to sit in with the personnel on guard in the main surveillance centre and follow their routines, sensory work and discussions.

Thus positioned with the Guardia Civil border guards, and after carefully avoiding any attempt to approach some of the many migrants who had made it across the fence and were staying in Ceuta, I was nevertheless contacted by the leader of a non-governmental organization (NGO) working with migrants. She had heard of my research from someone collaborating with the border guards and asked me to do a talk about my research on borders. A large group of sub-Saharan migrants came to hear my presentation. They, unlike myself, had very intimate first-hand experiences of that particular border fence and, as I soon discovered, they also had a very sophisticated knowledge of the border technologies deployed along it, which they readily shared (see Chapter 4). Talking to me about their technological expertise and their achievements in defeating the fence – and in French, a language they spoke well – clearly put their proficiencies to the forefront and thus opened a rich field of border knowledge, experience and organizational craftsmanship.

This dual position did not pose a problem for anyone. In Ceuta, *everyone* was entangled in the border from many different simultaneous perspectives: an NGO leader who was married to a chief of police, a Red Cross worker who was the son of a military commander, a former undocumented migrant whose brother was now a border guard and so on. My engagement with both border guards and migrants was therefore not an exception but, rather, the rule. Thus, I spent my fieldwork discussing the border situation and the different technologies with both migrants who had made it across the border and with the border guards whose job it was to stop them.

Enskilment and learning

Both chapters that follow focus on the interaction between technologies and human senses and on how they configure one another. An important aspect of the analysis concerns the enskilment of the senses, notably how border guards learn to see, both directly and through images, screens and visual technologies, in general. Seeing is not an inborn capacity but is socially and culturally acquired (Grasseni 2007b; Okely 2001) through a community of practice (Grasseni 2007a). In the case of the border guard forces with whom I worked, visual and sensory enskilment mostly came about informally, by working together and sharing tricks and experiences, rather than through instruction. In Copenhagen Airport, a unit specialized in document fraud presented new examples of forged and new, allegedly forgery-proof ID documents each week, teaching the border guard team how to identify the visual and haptic signs of both (see Chapter 3). In a training program for civil border guards, established to satisfy an urgent need for more border guards since the introduction of increased border controls in 2015, the trainees were taught to visually scan the perimeter by looking 'down, up, out – down, up, out', that is, down at the document, up at the face and out at people queuing in the border zone.

Thus, there were particular ways of seeing and sensing that required a specialized enskilment of the senses, as we shall see in the following chapters. However, the border guards also developed their own individual skills based on their own personal experiences and on special senses in which each of them excelled. Many explained how their 'sixth sense' and intuition were indispensable in carrying out their work, often in comparison with and in criticism of the machines that could not muster such imaginative and pre-emptive faculties (see also Møhl forthcoming). And whereas many of the details involved in carrying out their tasks could be taught through organized and collective training, the elements of their tacit knowledge, such as intuition and perceptive tricks, could not be directly passed on but could only be acquired through prolonged experience. The community of practice that involved learning to see also comprised the technologies and the human-technological interaction – for example, learning to 'see like the machine' (see Chapter 3), as well as teaching machines and algorithms what to look for (see also Part I). Thus, my work with this expanded community of practice consisted in picking up elements of both technological, shared and individual skills deployed in the sensory work of border control.

Not surprisingly, my own seeing was also gradually becoming enskilled throughout my fieldwork, although my immediate field of vision and sensing was obviously larger, including also the guards themselves and the technologies, as well as the ways in which the political and organizational background manifested itself minutely in the daily border work and decision-making. I was, for once, not working with a camera of my own, with the particular temporal and spatial selectivity that this implies (Møhl 2011; Møhl and Hauge Kristensen 2018) but was just as focused on detail, interaction and spatial configurations. In addition, I was observing how people and technologies were seeing interactively and the kinds of details they were occupied with, applying a kind of 'double vision'. However, when sitting in with the border guards and looking at screens and at the people, faces and documents passing by, discussing possible matches and mismatches, I was simultaneously doing the same interpretive assessment work as they, learning from them to look for signs and, like them, relying on my own personal visual experience. In this respect, I shared the experience of the new recruits, gradually learning to discern and distinguish the significant from the insignificant (see Chapter 3). As one civil guard said, she saw a threat in 90% of the passengers when she started but was now, nine months later, down to 10%. In other words, through her own experience and collegial advice, she had found a balance between heightened vigilance and getting the work done.

How images make sense and go between

Most biometric technologies function on the basis of photographic imagery, whether as modes of recording and verification, as in fingerprints and facial recognition, or as processes of human interpretation, as in surveillance imagery and bone scans. Such biometric procedures of verification are intrinsically linked to and contingent on what are assumed to be the trustworthy and evidential qualities of the photographic image. The notion that a photographic image can provide evidence, and even legal proof, lies in a combination of two semiotic features: first, its convincing visible accuracy – in Peircean semiotic terms, its iconicity or likeness to the object it represents – and, second, its indexicality, that is, its quality as a (photosensitive) imprint of something in front of the lens, providing it with a direct causal or 'physical connection' – a 'contiguity' – with its object (Kang 2014; Peirce 1998; Pinney 2008; Winston and Tsang 2009). In this sense, it obviously has a parallel in the fingerprint that carries validity both because it results from the direct imprint of a particular finger on paper or a scanner and because it bears a likeness to the original fingerprint when read and interpreted by both human and algorithmic eyes. The same goes for facial recognition: whether a human or an algorithm does the work of recognition by looking at and comparing an ID photo to a face, it is a likeness that is sought after. But the foundational authority of the process, recognized by all the institutions authorizing the validity of an ID photo, lies in the ID photo's indexicality, that is, in the idea that it is intrinsically connected to *one particular face* and, by extension, to the data connected to that face.

The chapters in Part II question some basic assumptions that we tend to take for granted about vision and imagery and thus about the 'vision work' that goes on in border control. Indeed, just as seeing is not a simple innate capacity, images are not simple representations or replicas of objects. As signs, they become objects in themselves, liberated from the dyadic relationship between an original and a representation. Here, the image is not just the representation of an original that it should reflect as faithfully as possible; instead, it becomes the main actor in the border process, as we shall see in Chapter 3. Yet, in this legislative setting, images are still seen as direct representations of those faces in a simple dyad and authenticating the relationship between a face/body and an ID, whether the reader is a human border guard or a facial recognition algorithm.

As objects in themselves, images also come to function as 'go-betweens in social transactions' (Mitchell 2002: 175). In border control, they act as go-betweens between humans as well as between travellers and algorithms, but they mainly function as go-betweens between faces and data sets that otherwise would not be connected. By 'objects in themselves', I am referring to their semiotic quality as separated from the objects they signify, not suggesting that they become independent ontological beings or agents or that they are imbued with intention or responsibility.

And finally, despite their indexically established authority, images evoke, allude and point out rather than simply tell, denote or explain (Berger 1982; MacDougall 1998). As Peirce writes, 'icons and indices assert nothing' (Peirce 1998 in: Chandler 2014); an image is reputedly worth a thousand words, but 'from any image countless possible statements could be inferred' (Chandler 2014: 132). Yet, 'as soon as photographs are used with words, they produce together an effect of certainty, even of dogmatic assertion' (Berger 1982: 91). Thus, where the photo is weak in meaning but carries validity because of its likeness and indexical quality, words acquire authority when they are attached to specific images. In terms of the identification and verification of identities based on ID photos in border control, it is thus the *connectedness* of the data set in the document and the *indexicality* of the ID photo that provide this last element of authority and 'dogmatic assertion', of 'knowing who someone *is*'.

The chapters

In the different border settings I visited, a variety of technologies and procedures were used, sometimes the same in very different locations, sometimes different ones within the same structure, every time involving new scales of human–technological interaction and sensory work. To acquire a better understanding of some of the fundamental processes involved, I have divided them into two overall categories of border control, respectively recognition and presence detection, presented in a chapter each, and defined respectively as 'hard' and 'soft' biometric technologies. Both types of border control are biometric in that they register and measure qualities of the body, but they work on very different technological, temporal, spatial and sense-making grounds. It is these differences, their basic functions and respective effects and their existential and semiotic/sense-making qualities that the two chapters dwell on and juxtapose.

Chapter 3 analyses the use of facial recognition in automated border control, a technology that verifies ID through visual analysis of facial traits. The chapter analyses and compares algorithmic and human procedures of visual recognition, as well as procedures for identifying threatening objects in luggage. It also analyses the role of imagery for purposes of identification, as well as processes of visual enskilment and deskilment.

Chapter 4 presents and analyses forms of presence detection in the control of border transgressions by using sonar and haptic technologies that 'listen' to and 'feel' for presences and hidden persons, as well as different forms of imagery, notably radar and infrared. They are considered 'soft' biometric technologies because they do not identify individuals but only kinds of bodies. The chapter finally inverses the perspective to encompass different forms of surveillance of the surveyors themselves, whether by migrants or management.

Comparing 'hard' and 'soft' technologies such as facial recognition and presence detection will also somewhat temper the widespread assumption, whether critical or celebratory, that digital technologies are more efficient and difficult to circumvent than simpler, 'softer' or older types of border control. The chapters serve to provide better and more detailed insights into the basic functions of each technology, the different types of encounters and narratives they entail and, as a consequence, some of the fundamental sensory-technological modi operandi of border control.

Note

1 The sea not only provides a 'natural border'; it is also politically enrolled to do the work of deterrence and, ultimately, the removal of threats, a strategy described as 'necropolitics' by Jason De León in reference to the US–Mexican border and the large stretches of desert to which many migrants succumb (De León 2015). See also Chapter 4.

3 Vision, faces, identities

Technologies of recognition

This chapter analyses technologies of border control that operate through *recognition*. The technologies are used in various border settings but are united by a shared techno-temporal quality, namely that they identify persons and objects by comparing them to *already-registered* IDs and *already-known* threats. The chapter first describes the principles and visual skills involved in running the Automated Border Control system in Copenhagen Airport, comparing them with face-to-face ('manual') border control and random checks on intra-Schengen flights. Second, the use of an X-ray scanner to determine the contents and possible threats of passenger luggage in Gibraltar Airport is analysed. In all these cases, humans and machines deploy refined and often interlaced visual and sensory skills or methods to determine whether a person or an object may pass or requires further scrutiny. This work engages with a complex combination of data, recognition and sensory skills that come to be mutually formative. The machinic operations are largely based on and formed according to human capacities and, at the same time, format how and what the humans operating the technologies come to sense and see, as well as 'unsee', in an interplay of continued enskilment *and* deskilment. We could tentatively characterize this integrated human-technological border assemblage as cyborgian (Haraway 2016; Wells 2014) in that human and technological processes and sensory connections supplement and reinforce one another. But, as we shall see, they sometimes also destabilize one another's operations and skills and work in more contested and disconnected ways, thus undermining the analytical pertinence of the connected symbiotic cyborg.

The chapter further analyses the kinds of *suspicious relations* – notably between faces and identities – that arise and undergo scrutiny in border control and the *thresholds of resemblance* that are required to pass border control: that is, how much one needs to resemble one's ID photo. Finally, the chapter presents a particular element of airport security control where known threats appear artificially on the security agents' screens to check not the luggage itself but the skills and alertness of the agent. In all these examples, sign relations are established between objects and meanings. On the surface, these relations appear simple and direct, but they are, in fact, enmeshed in and enskilled through a complex political, economic and social community of practice. The chapter attempts to single out a few of these semiotic operations, the forces they are regulated and influenced by and the effects they have on the production of identities, threats and borders.

Settings and technologies

As mentioned in the introduction to Part II, I carried out fieldwork with border police in three main settings: Copenhagen Airport, Gibraltar Airport and the territory's land border with Spain and, in Ceuta, on the border between Spain and Morocco. The goal was to analyse the different types of technologies that were used to detect and deter threats and to compare their workings. The border control processes I present in Chapter 4 concern different forms of *presence detection* used mainly in Ceuta. In the two airport settings, however, control was, to a large extent, technologized and automatized, and control of identities and objects took place through processes of identification and recognition, as we shall see.

This chapter therefore takes us to the two airport settings, starting with the Schengen passport control in Copenhagen Airport. Here, police and civil border guards supervise the running of an automated facial recognition system devised and installed by the Portuguese company Vision-Box and maintained by a Danish service company, Biometric Solutions. The system is called the ABC – Automated Border Control – and consists of nine 'eGates', three of which can be adjusted for use from both sides to accommodate different passenger flows. Besides the automated eGates, the border zone also consists of a row of 'manual' passport control booths where border guards physically check passports and visas. Specifically, for migration control objectives, police officers also carry out random checks on intra-Schengen flights arriving from so-called high-risk cities in Southern Europe. After describing and analysing the sensory and interpretive work involved in these operations, the analysis then moves to Gibraltar Airport, where police officers survey the screens of X-ray scanners, inspect passenger luggage, detect possible threats, make assumptions about the identities of the travellers from the objects they can identify and assess whether it 'all makes sense'.

In all these cases, the border guards are presented with imagery, scans, objects and faces that they need to single out, identify and assess in a series of complex human–machine interactions and intricate processes of reading and interpreting signs. These readings are based on already-identified threats, as well as on assumptions about the translucence and authority of the images and their direct relation to specific persons, objects and lives.

The analysis in this chapter is based on fieldwork conducted in Copenhagen Airport and Gibraltar Airport over a period of five months during 2016–2017.

Facial recognition and looking like a picture

At Copenhagen Airport, Hanne, a border guard, is surveying her two screens. She has turned on the computer and the ABC system that monitors the eGates, has logged into the system and has opened the access to the different national and international police databases (national ID, watch lists, wanted persons, etc.). She activates the six eGates she is in charge of, and the green lamps turn on, inviting the first travellers into the ABC. From her little glass booth, Hanne has an overview of all the gates and the travellers passing through them. She picks up her mouse and starts working, shifting her attention between the ABC screen, the faces and bodies of the travellers, and the database screen when something appears on it. With a click, or using the touchscreen, she can 'help' the ABC, checking the irregularities it identifies, overriding it when it makes mistakes and turning it off when it 'starts glitching'. Otherwise, she simply checks that it 'does its job'. She is in radio contact with her colleagues on the floor and in the other control booths and can tell them when she sees something suspicious – people changing queues or looking nervous – or switching from the ABC to the manual control booths.

The ABC uses a facial recognition system where the ID photo in the biometric passport is compared to the face of the person passing through the control. The passport contains a so-called template of the facial features of the ID photo, that is, a series of measurements of distances of key points in the photographed face. In passing through the eGate, the same features are measured on the traveller, and a new template is produced and compared to that from the ID photo. Thus, contrary to human recognition, algorithmic recognition does not compare actual faces to photographed faces but templates extracted from both. Comparing the template from the photo and the one from the live face, the ABC algorithm produces a comparison score that has to meet a certain threshold for the exit door to open. The score can range from 100 – total resemblance – to 0 – no resemblance whatsoever. And it actually does vary quite a lot while Hanne is looking at her screen.

Figure 3.1 Comparing 2D photos to 3D faces in the ABC booth
Source: Photo by the author.

Hanne is a senior police officer who has been working in the ABC since it arrived five or six months previously. She is a 'super-user', training newcomers, and she likes working with the ABC, defending 'it' vehemently when a colleague launches into criticism of it during a break or when I ask her. She finds it amusing, even touching, when it mistakes a suitcase for a second person, when it cannot see because of the sunlight or when 'it gets confused' and starts glitching. She talks to it and often laughs, describing it as having a consciousness, moods and sensory frailties that resemble her own, as when humans ascribe consciousness and will to their computer to explain its intricacies and malfunctions (Jackson 2002). To her colleagues' objections that in the end it will make them all useless, she says, 'Well, it can be turned off. We can't'. She is confident that human skills will always be indispensable in border control. She will be retiring from the service in a couple of months, a year at the most – which might explain why she does not feel menaced by the machine as some of her younger colleagues do.

A traveller arrives at the entrance to the eGate and places his biometric passport on the scanner, and the passport is read by the ABC system. After some moments, the passport photo and the chip photo, as well as the contained ID information, appear on Hanne's screen. If the two photos are identical – that is, if the picture in the passport has not been replaced with another one – and if the passport is biometric, undamaged and in compliance with the ABC system, the first glass door automatically opens, and the traveller is invited into the eGate. Until now, only the passport has undergone scrutiny. Now it is time to check that the person in the document and the person in the eGate are the same – or, rather, that the face in the document and in the eGate look sufficiently alike.

The traveller is asked to stand on a pair of yellow footsteps on the floor, the entrance door closes and the traveller is trapped inside the eGate with no possible escape route. The camera moves up and down, small lamps blinking, and finally settles in front of the traveller's face. What goes on in the entrails of the ABC – what it sees; what it is doing, thinking or processing; which databases it is linked to and checking right now, what it remembers – all that is invisible and unknown to the traveller, who just waits in front of the camera and the closed door. From her position in the ABC booth, Hanne can check the ABC's algorithmic operations on her screen by making a line of algorithmic operations appear in a little window next to the facial images of the traveller. Prompted by my question, she opens the operations list and admits she has never really looked at it before and does not know how to read it. She has up to six gates to survey and no time to check what the ABC is doing, as long as it is running smoothly and showing no alerts of detected irregularities or 'signs of fatigue', as she says.

The live face of a traveller in the eGate shows up on Hanne's screen, and the ABC starts comparing the facial templates from the passport with the direct cameras. If the comparison score is high enough, meeting the 'recognition threshold' – usually somewhere between 40 and 50, depending on the settings – the gate will open, and the person has successfully passed the border. But it does not always go that smoothly. The ABC system is hyper-sensitive, Hanne says, and if one eGate starts having problems, it sometimes spreads from one gate to another 'like a viral infection'. If there is too much light or too little, or if people are moving around or are carrying a lot of bags, it 'can't see properly'. Often a second-line police officer has to step in to indicate the right position and posture if the traveller does not comply with the system's specifications. 'Stand still, not too close'. 'Take off your glasses'. 'Take off your hat'. 'Remove your veil'. 'Look at the camera' – modifying their appearance until the door opens.

Thus, in fact, to pass through the ABC requires that one looks sufficiently like the face in the ID photo, whether the passport is one's own or someone else's. The little ABC enclosure becomes a theatre of inquisition where one is under suspicion until the algorithm has finished analysing and comparing faces and templates, and the glass doors open. On this stage, people are told straighten their hair, smile or stop smiling, in order to 'look sufficiently' like the photo to meet the threshold of resemblance. The ABC is reacting directly to their capacity to look like the face in the photo, displaying the changing comparison score – . . . 13 . . . 35 . . . 22 . . . 44! – as the face in front of the camera changes its appearance. The ID photo itself has been formatted for the occasion of making the passport, with strict instructions on how to look and how not to look.[1] More than a site of

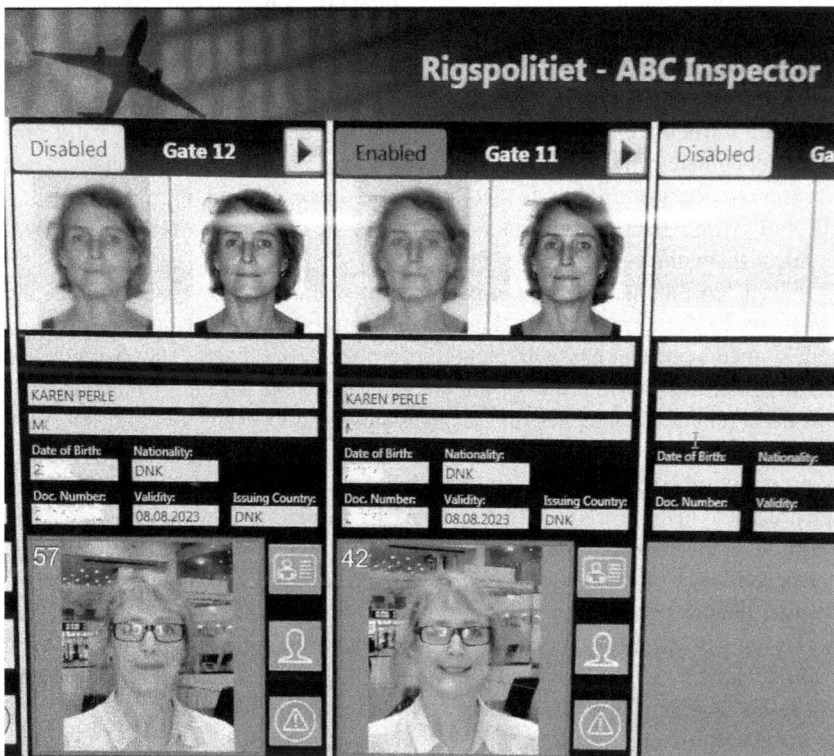

Figure 3.2 The author in the ABC, trying to look like the ID photo, more or less successfully

Source: Photo by the author.

identification, the ABC becomes a scene of semblance, of mimicry, of creating an iconic likeness with the photo – a photo that is an already-formatted version of a face, often captured many years earlier. Thus, the recognition system, whether automated or human, is based on the comparison between a supposed original – which here ironically becomes the ID photo – and a suspicious-until-verified replica, the face of the traveller.

Inversely, what the ABC and the border guards are assessing is not the identity of a person in all his/her social and existential intricacies – something that could never be transformed into simple ID information (Feldman 2013; Møhl forthcoming) – but the travellers' skills in iconicity and mimicry of a small 2D photographic version of a face. All authority is vested in this small simulacrum and in its indexical relationship to a particular ID, and control consists in the capacity of the algorithms and the border guards to see and read it, to recognize the ID face in the living face. One could argue that the ID photo has taken over the identity – or at least the ID – of the person who was originally photographed. The photo has become the original to which the traveller must conform, whether it is his or her own or someone else's passport. As Susan Coutin observes, to pass a border, one becomes eligible by producing the adequate documents, for 'documents confer, rather than derive from, statuses' (Coutin 2003: 60). A person can 'claim an identity' simply by waving a document, as a frustrated border official remarked to Gregory Feldman (2013: 142). As such, the simplest way to derive an adequate status and pass the border is, Coutin notes, to purchase fraudulent documents. Or, since the introduction of biometric passports in Europe, to purchase valid biometric documents with ID photos with which one can produce a reasonable level of resemblance, just as long as there is a valid document and, in my analysis, an ID photo with which to conform. As such, it is the capacity for resemblance and mimicry that confers status.

Indeed, documents *can* be tampered with; photos *can* be replaced. But the DocuUnit at the airport police department teaches the border guards to detect forged documents and passports, and they are pretty good at it. They learn to see the irregularities in the printing, the cuts and tears, the missing watermarks; to feel the stitching, the thickness of the pages and the embossing; and to listen to the sound of the plastic when tapping it with a nail or on the desk. The differences are minute and imperceptible to the untrained, but the border guards' enskilled senses are proficient in these matters. And in the ABC, the algorithm compares the visible ID photo to the chip photo to determine if they are identical. So, forged documents are becoming a rarity, especially since the introduction of biometric passports. Instead, the most common document fraud currently detected in Copenhagen Airport is not persons travelling on forged documents but, rather, on another person's *valid* document – 'impersonation' (Frontex 2013). This is confirmed notably by our research among migrants (see Chapters 6–7) who prefer to travel on borrowed or bought legitimate documents when trying to pass border control. All it takes is to look sufficiently like the person in the small ID photo.

On this little stage of mimicry and unmasking, the border guards employ their own personal tactics. Hanne has been working with the ABC since it was installed and is gradually adjusting her senses to the work of the machine. As she says, 'I'm learning to see in 2D', comparing the 2D screen photo to the actual '3D' face in the eGate. She enjoys trying to outsmart the machine, 'seeing more quickly' than the machine, making her identity assessment before the ABC opens its doors and sets the traveller free. And to do that, she says, she has to 'see like the machine'. According to this logic, her vision is being transformed by adapting it to the ABC's vision and scans of small ID photos, switching between the depth of living, changeable, fleshy faces in front of her and the black-and-white imprints of their superficies. She is *learning to see* in *other ways*; her vision is being formatted by how the ABC sees and by the materiality of photographic technologies and how they work and convey knowledge – how they epitomize and essentialize identities by linking those fleshy faces and bodies to the data in the document.

I ask Hanne if she thinks the ABC sees better and is more efficient than humans. She is not sure, she says. 'No one is foolproof, not the machines nor the humans'. She alludes to the predicament that no one can know how many have actually managed to fool passport control. 'How could we know? They went unnoticed!' Then, to make her point, she tells me that some days earlier, an airline agent prevented a traveller from boarding a flight because he was trying to travel on another person's passport. Yet he had made it through the manual passport control. The officers of the DocuUnit apprehended him and, while they were at it, ran him and the passport through the ABC three times to see how it reacted, and he actually almost passed once. The case speaks directly to the overarching predicament of nescience in policing and security: never knowing who was not stopped, how many went unnoticed. But Hanne keeps the discussion at a more practical level: 'People who want to cheat probably don't choose the ABC. I'm not sure it works better, but apparently people think it does, so they chose the human border guard . . .'. She concludes, 'It depends on a lot of things. Sometimes the machine sees better, sometimes the human eye does'.

Indeed, when I have seen cases of obviously different faces in the ABC and in the scanned passport, it has always been someone who accidentally took the wrong passport, for example, couples inadvertently switching passports just before going into the ABC or, as happened once, someone who took his mother's passport with him or a border guard checking that the ABC is up and working, as one of them does every morning when he starts his guard by trying to pass through the ABC using a prop passport and grimacing to mimic the ID photo of a rather chubby man. Through this routine check, he also seems to be assuring himself that the ABC – and, with it, the whole service – cannot be fooled, that it is still sufficiently on guard and, importantly, that the recognition threshold has not been lowered overnight. In sum, he checks that the border is still there and working.

In the ABC, ID photos play a major part in the work of border control. The ABC itself 'looks at' and compares faces with ID photos, the border guards in their booth do much the same, keeping pace with the machine, and everything is, to a large extent, based on the indexical authority of the photographic image to contain and convey the truthful link between face and identity.

Figure 3.3 When the ABC starts glitching, there are several solutions; one of them is to turn it off

Source: Photo by the author.

Hanne's enskilled vision plays a huge role. But she also uses her other senses, listening to the sounds in the room and on the radio, surveying the bustle and the reactions of people in the queues. This multi-sensoriality becomes even more acute when she moves over to the manual passport control.

'Manual control', making sense and questions of plausibility

In both the ABC and the manual control booth, the border guards use their eyes to look at passports, photos, faces and behaviour: who is looking nervous, changing queues and so on. But in the 'manual' control, the border guards sit close to the travellers and deploy '*all* their senses, including the sixth', as Ole, one of Hanne's colleagues says. And in the manual control, Ole adds, they are not restricted, as in the ABC, to controlling travellers on the basis of already-registered images and data. In the manual control, they can pry into the future; they can use their 'creativity and intuition'. 'You can have people whose papers are perfectly in order, but who behave strangely'.

I sit in with Ole and Hanne in their joint manual control booth in a border zone where flights arrive from extra-Schengen airports defined as 'high-risk'. Arriving passengers are separated from other passengers by an intricate network of corridors that sends them to an extra security check and a secluded baggage reclaim because they are arriving from airports where security control is considered insufficient by Schengen standards.

Ole and Hanne comment on their work as the travellers arrive, giving a sense of the variety of their sensory work. They check heights using an old-fashioned height-meter. They look – down at the document, up at the face, out at the queues: down, up, out. They listen to voices, explanations, intonations and accents. They touch the passports and feel the stamps, the stitching and embossing, the new-nesses and the wear-and-tear. They smell smells. They ask questions: 'Where are you going?' 'What is your purpose there?' and 'Who are you visiting?' They chat about destinations and origins, about the travellers' children, about the weather. They scan passports, fingerprints and make photocopies. And while they are doing all that, in the 20 seconds or less that they have at their disposal for each traveller, they stitch together all the minute details they pick up, weaving the fragments into stories about the travellers' intentions and agendas and assessing the plausibility of those stories and whether they make sense or not, or, as Ole and Hanne put it, 'whether two and two make four – or five'. They are pre-empting, thinking ahead, imagining and assessing potential threats – that is, events in the future – based on the thickness of the here and now and not only on sparse ID data from the past and a small ID photo. Unlike in the ABC booth, they work with their versatility, their capacity to cognize, to improvise and to combine disparate details and imagine scenarios about unknown futures. Ole definitely prefers the manual control and the human contact it requires: 'The machine can't think. It's stupid and can only obey orders given by humans. It can't think ahead. I have intuition, I'm creative'.

The way they look at people here is different, 'more full-scale', as some of them say. This is because they deploy *all* their senses in a synaesthetic manner where the senses collaborate to thicken the appreciation and perception, and because in the direct interaction, they can 'read' people in more dimensions: temporally, with regard to their past movements and their intentions, and socially, in their interactions with others and in the ways they perform and present themselves. The enskilment of the border guards' senses is not, as in the ABC, formatted by a particular technology that they learn to see through and that they also have to watch over, surveying its modes of vision, its visual impairments and its viral infections.

Analysing vision and sensorial-aesthetic relations with the farming landscape in Normandy, Judith Okely distinguishes between a distant, unengaged gaze and a seeing that involves the entire body and all the senses, and that is based on years of practice and physical bodily engagement with the landscape (Okely 2001). When Okely looks at the agricultural landscape, she searches for 'significance in everything encountered' (ibid.: 111), as I do when I look at luggage scans and travellers, and search for meaning in every sign. She and I have not yet been trained to focus selectively on particular elements. In border control, as in farming – as in every 'community of practice' (Grasseni 2007a: 203) – there are particular ways of seeing and identifying significance. In border control, training programs are set up to teach the border guards and security officers *what to see* and *what not to see*, notably when looking for details in a passport and scrutinizing a face. Cristina Grasseni, working with cattle breeders who really know how to look at a cow, operates with the same fundamental distinction between distant and immersed forms of vision. She describes 'good looking' as a trained perception, a cognitive form of apprenticeship that is culturally inculcated and socially performed (Grasseni 2007b), a schooling of the eye that 'is at once aesthetic, moral, functional, and normative' (Grasseni 2018).

The political and ideological background of border control and the way the particular border is defined and determines threats and intruders form a basic setting for all the border guards' work, whether in the ABC or in manual border control. But the technological, material and sensory aspects of the different work stations and tasks seem to play just as important a role in the way the guards perform their tasks and how they define and detect threats.

In the following, we accompany a couple of border guards on another type of 'manual' mission in order to examine how it is mainly the *organizational* framework of that mission that makes the border guards *see* in yet other ways and focus on other objects, aspects and appearances than they do in the ABC and the manual control booth.

'Random checks' on 'high-risk flights': looking like a suspect

Border police officers in Copenhagen Airport carry out a certain number of random ID checks every day on flights arriving in Copenhagen from other Schengen countries, which are normally exempt from border control. These checks are audited by police management, both to monitor that the checks are being done and to document the number of weekly checks in case Frontex, the European border control agency, comes to control their activities.[2]

Walking to the airport terminal used by budget airlines, where we will await the arrival of a plane from an Italian city, I accompany Hans, a short and sturdily built police agent, and his new female colleague, who is carrying out this type of arrival check for the first time. Hans explains the principles to her as we walk: 'We're looking for illegal immigrants. So we don't stop "Mr and Mrs Hansen"'. She nods and seems to know what she is required to look for. While we wait for the plane to dock, I ask Hans how he can see whom to pick out if he cannot stop everyone. He has some techniques, he says. He can, for example, detect undocumented Afghan men without checking their passports: 'They try to blend in but always look sort of inept-smart. You know, jeans with holes in them. Sunglasses. And there's always something with the shoes – they never fit the rest. Too fancy'.

The flight arrives, and they take up position in front of the arrival gate. Most passengers seem to be Danish tourists coming home. But four men, arriving one by one, could be of African origin. The first is pulled aside by Hans and, having observed him, his colleague pulls aside the next one and asks him to produce his documents. None of them put up any resistance. A third man is stopped by Hans, and a fourth passes by while the officers are preoccupied with the others. All the ones they check have valid documents. Hans smiles jovially and wishes them a welcome home or a continued good trip. He notes down the flight and number of the checked persons on a piece of old wrapping paper – 'So the officer on duty can enter it into the system' – and the two walk off to the other end of the airport to 'randomly' check another flight.

Thus, what sorts of factors play a role in and frame what Hans and his colleague see and how they make distinctions in the task they are being asked to perform? Rather than personal choices and prejudices, it seems to be the notion of 'high-risk flights' that configures which persons Hans decides need to be checked and which persons to let pass. 'High risk' is used in a double sense by Frontex: (1) as a way to optimize resources by concentrating them on flights arriving from European regions of high migration influx (European Commission 2017a), instead of indiscriminately checking every flight – that is, in an organizational sense – and (2) when referring to passengers' specific places of origin, that is, in a geographical sense. At Copenhagen Airport, certain flights are therefore defined as high risk because they arrive from cities within Schengen that Frontex and the national authorities consider possible avenues of 'illegal migrants' and refugees from the Middle East, Asia and Africa. When Hans and his colleagues randomly control such flights, the notion of 'high risk' therefore comes to relate to the countries of origin of the possible migrants and asylum-seekers, meaning that they are, in fact, profiling nationalities and not just checking 'randomly'. The notion of 'high risk' therefore logically, including for the border police officers themselves, implies that they are required to check 'people who *look* high-risk', which in this case means 'from Africa or the Middle East'. And, as several officers have said to me, this is paradoxical because it makes them take decisions based on what effectively amounts to 'ethnic profiling'.[3] And even if they might want to contest the legality and ethics of these random checks that are making them select people based on how they look, the border control unit needs to do a certain number of random checks every week for future Frontex controls. As one border guard said, 'If I wasn't a racist before, this work is turning me into one'. Her personal opinion of certain ethnic groups had not changed, she said, but she was de facto required to be more suspicious of certain categories of passengers than of others based on their external physical appearance. She was being asked to apply a synoptic 'tunnel vision' and to 'see like a state' (Scott 1998: 11) – in this case, a union of states, the EU – and not like a human being.

In this example, the technologies involved are not material in kind but infrastructural, involving a range of EU policies and management orders, as well as the travel routes and modalities of migrants, which all contribute to establishing a particular form of legibility (ibid.) and set the conditions for how the border guards deploy their senses, what they look at and how they make decisions based on their visual perceptions. A pair of shoes that are too fancy or a particular kind of face or skin colour will raise the border guard's perceptions above the suspicion threshold and provide the incentive to control the traveller.

In a final example of recognitive skilfulness from my fieldwork at the Gibraltar Airport, we will see how border police use X-ray scanners to look through the surfaces of travellers' luggage in search of threats. This example demonstrates how they are required to both *see* and *unsee* and how their vision is being both skilled *and* deskilled by a mesmerizing system that is intended to keep them awake and alert to threats – specific *known* threats, that is.

Luggage scans, Threat Image Projections and the deskilment of vision

In a secluded room at the bottom of the Gibraltar Airport terminal building, Veronica, a police officer, is surveying her screen. This is Security Level 2. There are no windows, and the artificial light is faint. She is sitting by a long desk, the only piece of furniture in the room except for a couple of chairs. She offers me one and seems content to have company. Somewhere else in the building, in Security Level 1, an automated X-ray scanner checks all luggage before it is sent down to the loading area. If the scanner identifies a suspicious object or an unidentifiable mass, it loads the piece of luggage onto a special conveyor belt, and an image of the piece of luggage pops up on the screen in the cellar. Here, Veronica has exactly 12 seconds to inspect the image. She can either validate the contents as nonthreatening by hitting the green button on the screen, sending the piece back into the ordinary flow of luggage headed for the airplane, or, if she sees a threat or has doubts, she will hit the red button and send it on to Security Level 3. If she does not make a validation within the 12 seconds, it is taken as an indicator that she is either in doubt or inattentive, and the machine automatically sends the luggage on to the next security level.

In another room, Security Level 3, in the midst of the conveyor belts, Veronica's colleague John receives the pieces of luggage that Veronica or the automated system have sent on for further inspection. John receives less luggage and has more time to look and reflect. He can turn the luggage X-ray image around, inspect it from all sides or move through the layers, and he has the time to see and identify particular objects to make his decision. But back in Level 2, Veronica only has her 12 seconds to look at the image and figure out why the automated system on Level 1 has deemed the piece suspect. She can use different filters to discern organic from non-organic and metal from soft materials, all displayed in different colours. But she rarely looks at what is inside the piece of luggage. She only has the time to register the blurry zones, the questionable unknowns, and to check for a couple of known threats – explosives, electric circuits, firearms – before hitting one of the buttons. She does not actually *see* what is in the piece of luggage, she says – at least in the sense of remembering or being able to list it, attaching meaning to it. She is incapable of telling me what was in the last suitcase when I ask her.

We both look intently at the screen searching for threats as the pieces of luggage pass by, one after the other. 'I have never seen an actual gun or bomb', Veronica says. But she knows they will show up once in a while and that she has to be on her toes. Indeed, at regular intervals – every 10 minutes or so – the machine produces a so-called TIP, a 'Threat Image Projection', an X-ray image of a known security threat projected onto a piece of luggage.[4] Veronica has to identify the threat and hit the red button within the 12 seconds. 'This is to keep me alert', she says, adding that it is also designed to check that she is doing her work properly and not sleeping. The results of her clicks on the TIP items are registered and presented to her at the end of the month and can lead to reprimands, she says.[5] After some minutes, a suitcase arrives with an unmistakable gun inside; she hits the red button, and the image disappears. 'You saw it?!', she asks. I did. There was something uncanny about it. But I could not say what, as it disappeared so quickly.

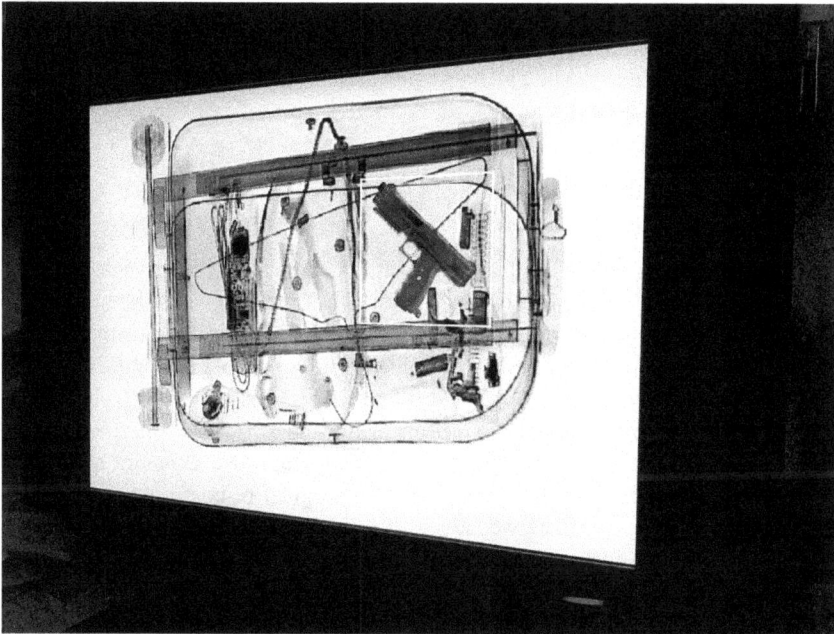

Figure 3.4 Example of a TIP image

The TIP system is, in principle, intended to counterbalance the routinization of the selective vision (Cutler and Paddock 2009) and a correlated unseeing because threats so rarely, if ever, appear in actual traveller bags and thus on the screen. It is therefore intended to ensure that officers are regularly exposed to threats – albeit imaginary threats – so that their vision may be enskilled to seeing those threats, despite the lack of direct perceptions of actual firearms, weapon parts or explosives.

This means that the threats that Veronica and her colleagues see are in fact not found in the travellers' suitcases, but are produced by algorithms, technology developers and their depictions of *already-known threats*. New and as-yet un-encountered dangerous objects cannot be detected by the algorithms in Level 1, nor can they be projected on to the luggage images to train the officers to see them. The TIP algorithm therefore potentially produces another type of *visual agnosia*, since it repeatedly projects only *knowns* on to the screens and into the sight of the officers, formatting, limiting and quite possibly *deskilling* their perceptions by training them to recognize only what is already known instead of using their own skills and intuitive sense of what could constitute possible threats.

In addition to its pre-emptive functions and its potential deskilment of the officers' senses, the TIP system, as Veronica also suggests, serves to audit the officers' performance by monitoring detected and missed instances of artificial threats. This obviously situates the TIP system within the wide range of control mechanisms that govern by numbers, to which operators in any work setting are increasingly exposed (Shore and Wright 2015; Strathern 2000). It also further strengthens the suggestion that technologies of surveillance are, if not efficient in the detection of threats to national security, then at least efficient in controlling the controllers (see also Møhl 2018).

Comparisons and conclusions

This chapter has provided four examples of systems of threat assessment that work with different types of technology and infrastructural protocols. In the control booths in Copenhagen Airport, some protocols are data- and algorithm-based, whereas others operate primarily through human sensory and cognitive proficiencies and storytelling skills. These different operations of respectively human and technological observational skills, assessing the known and perceiving the unknown, are in practice tightly interwoven. They do, however, point to significant differences in temporality: between technological re-cognition and human cognition and, respectively, between identifying known and registered threats and felons based on data and prior registration and seeking out and imaging unknown ones based primarily on the ambiguities of encounters in the here and now. Most human sensory and mental activity is based on recognition, and the balance between perceiving knowns and discovering unknowns is finely tuned and mainly reflects different forms of human cognitive behaviour. The border guards nevertheless constantly acquire new selective tacit skills from their direct experience and from digging further into suspect or unknown objects and details to substantiate or refute their intuitions and thus broaden their individual skills and sometimes skills they share collectively. The data provided in the control booth and the TIP system, on the other hand, pointedly demonstrate the incapacity of algorithmic foresight. Both need the intervention of human agents to pick up and connect disparate details, to foresee and imagine connections and future scenarios. In that sense, data are quite impotent when it comes to assessing the now and the future.

The ABC system of facial recognition functions within a realm of semblance and simulacra. It places the process of identification solely in the face and in the 'photographic original' with which it is compared. The floor officer giving instructions is asking the person to conform to the photo by taking off their glasses, their veil, their smile. The person – and his or her identity – is reduced to, and implied to be contained in, the face, especially in the photographic version of the original face to which the traveller must conform to pass. There is an unfathomable faith in the indexicality of the image and its direct link to a previously existing face. In identity control, however, the image becomes the original and the traveller's face becomes the re-presentation in an ironic semiotic inversion of object and representation.

In all cases, trying to look like a photo or like a casually dressed young European – as Hans suggests some young migrants do, confirmed by the descriptions of migrants *en route* in Part III – amounts to conforming with and in a sense inhabiting a category (Rapport 2013). And, maybe most important, this points to the significance of the border guards' awareness that people can and do move between categories and that identity is not something fixed, as the logics of ID and data would imply.

The logics of the identification process also point to a hyper-focalization of the face itself, whether original or actual, as the locus of identity. The face becomes a hyperbole, as implied in the notion of 'faciality' (Deleuze and Guattari 1987; Gates 2011) and 'the imposition on the subject to assume a 'face'' (Rodrigues 2012: 3). In that sense, the ABC is a perfect example of what Deleuze and Guattari call an 'abstract machine of faciality' – it no longer works with humans or live faces or in any way with complex human identities but with abstract 'phantomized' template versions of both the photographed 'original' (linked to a data set) and actual faces template-fied in the ABC.

In the ABC and the luggage scans, the technologies tend to pick out zones and objects of interest and to produce threats, for example, as TIPs and as hits, productions that format the human senses and that may produce certain forms of agnosia in the interaction between these different forms of human–technological vigilance. By 'providing more exposure to threats', as the developers say (see note 5), the TIP image ensures that officers are regularly exposed to imaginary threat so that their vision may be trained despite the lack of direct perceptions of actual weapon parts or explosives. This means that threats are not constituted by travellers and their objects but are produced by algorithms, technology developers and their assessments of *currently known* threats. The threat projections thus operate within the domain of hindsight. New and unknown, as-yet unencountered dangerous objects can neither be detected by the algorithms nor projected onto the luggage images and trained by the officers.

Border control is governed by a fundamental predicament, a pervasive uncertainty that cannot be represented in numbers and audits because, by definition, it goes unregistered. How many persons have actually managed to pass control without the proper documentation? How many went 'under the radar'? There is no way of knowing. This prevailing *spectre of uncertainty* in border and security control sets a perfect scene for the prolific production of social imaginaries and storytelling. As we have seen, these stories are produced by assembling a variety of small cues into narratives that explain the presence and concurrence of those cues and that spell out apparent incongruities (see also Holmberg 2003). But they in no way create a full picture of who a traveller might be or have in mind – that dream of total transparency that any border guard system would like to possess and that the biometric technologies allegedly provide. Indeed, it is widely acknowledged that the technologies cannot keep that promise and that the small details are only fragments that provide a basis only for hunches, not for truths. Indeed, the great unknown prevails and seems to fill the room around the working officers as an underlying tension in every act of decision-making they engage in: Am I making the right choice? Is the ABC making the right decision? Was that person who the passport said she was? Was that strange object in the suitcase a kitchen utensil or part of a bomb? Is the ABC working properly?

The technological infrastructure that is supposed to help the officers in their work and eliminate most of the uncertainty tends to do the exact opposite, at least from the border guard's perspective. Take, for example, the indecisiveness of the ABC. Every time it shifts from one recognition score to another, swinging back and forth between degrees of likeness and recognition, gradually closing in on a passable recognition score and then lowering it again . . . – . . . 13 . . . 35 . . . 22 . . . 44! – it very explicitly signals to the border guard looking at the screen that *it is not sure.* It also signals that likeness is a very relative and negotiable matter, a question of degree and approximation – very rarely at 100, more often oscillating between 20 and 50. So where the door to the traveller is decidedly either closed or open – either/or – the displayed, colourfully shifting recognition score is wholly ambiguous and only underscores the uncertainty of the border control endeavour. In the same manner – besides deskilling the officers' vision – every time the TIP images appear on an officer's screen, they indicate and remind her that she and her vigilance are being controlled and are under suspicion. It places management control, suspicion and thus uncertainty at the forefront of the activity.

And basically, can we trust the ABC and its complete reliance on ID photos and likenesses? As an airline officer recently answered when I asked him what he had checked in my passport, 'I just check the date of expiration and the name'. I persisted, 'And what about the ID photo?!' 'Oh, I don't look at that. Photos can cheat'. Although facial recognition involves complex and very high-tech operations and calculations and is presented and merchandized as a new generation of biometrics – transcending both human capacities and older forms of identity verification and thus being the ultimate high-tech solution to the quest for shatterproof border control – facial recognition can, in fact, be circumvented with sufficient practical skill (or economic sway). The belief in its foolproof qualities belongs to the myth of the 'digital sublime' (Mosco 2004) and its claims to produce definitive truths about the relationship between faces and identities (Gates 2011) rather than radically different mechanisms of sign analysis and control.

While we should not belittle the potency of digital and data-based surveillance and border control technologies such as the ABC, we should not exaggerate it either. This chapter has, among other things, tried to examine the efficiency of a surveillance technology celebrated – or criticized, depending on the perspective – for its efficiency and capacity to create 'smarter', impenetrable borders. But, echoing Kelly Gates (2011) and William Bogard (2006), if we consider the all-powerfulness of surveillance and biometric technologies as givens, as 'already here to stay' – indeed, engaging ourselves in what Bogard calls 'the imaginary of surveillant control' – or if we succumb to the naïve fascination of the digital sublime without also acknowledging and analysing the shortcomings and sense-making bases of such technologies, we risk contributing to the illusion of their proficiencies. The myth, with its preventive effect, becomes part of the 'fortress'. Finding a middle way by paying greater attention to the concrete processes and authoritative, semiotic premises of such technologies, especially alongside their daily users, can give us a more fine-tuned sense of their workings and of their failures to fulfil that illusion of omnipotent, all-powerful surveillance and impenetrable control. What is interesting is to figure out the bases for the claims to the veracity of such technologies and the ambiguities that in practice they come to operate with and under – in this case, that it suffices to resemble a tiny photo of a face in order to pass a border.

Notes

1 See, for example, Danish police authorities' guidelines for ID photos concerning pos-tures, props and clothing: https://politi.dk/pas/krav-til-pas-og-koerekortfoto.
2 For a description of Frontex operations, see the general introduction to the present book.
3 Profiling based especially on ethnicity is formally prohibited by European law (European Commission 2000; European Union Agency for Fundamental Rights 2010).
4 'TIP' images can be either virtual threat items projected into X-ray images of real passenger bags or virtual X-ray images of whole bags (Schwaninger and Hofer 2005), depending on the type of screening system. In Gibraltar, both are used, depending on the security level.
5 According to the producer of the system, 'TIP is designed to advance screener proficiency by providing more exposure to threats on a regular basis, and to track screener performance' (Rapiscan 2017).

4 'Is it a donkey?' Presences, senses and figuration in human–technological border control

Biometric technologies and 'smart borders' are generally associated with complex, high-tech automated systems that control IDs on the basis of recognition (European Commission 2014; Sanchez del Rio et al. 2016), like the facial- and threat-recognition systems described in the previous chapter. Yet, as we shall see, simpler, often analogue technologies that merely detect bodily presences can be just as efficient, if not more so, in deterring supposed threats as identification and recognition technologies. Indeed, infrared cameras, radars, haptic sensors and sound detectors can identify the presence of a something or someone and qualify their kind but without identifying them as individuals. They are place-oriented and can detect signs of life and other forms of presences, substances, heat emissions and sounds, and they can also to some extent qualify that presence. If they compare, they do so by detecting likenesses to known categories – is it a donkey, a bird, a human? – but not to known individuals. The links they establish are generally between a body and a place, and it is that link, that *presence*, that can define the body as a threat, an intrusion – a body in the *wrong* place. As some providers say, they perform 'situational awareness' (Dynetics 2018). The selective intelligence work becomes to determine the *quality* of the presence and, if the presence is defined as a threat, to deter or apprehend the intruding body.

This chapter describes and analyses the workings of a series of such biometric presence-detection technologies as they are used to control border-transgression attempts in Ceuta, Spain, by using different kinds of visual, sonar and haptic technologies that 'see', 'listen' and 'feel' to detect presences and hidden persons. The chapter further analyses these technological practices of border work and the subjects they produce by comparing them to a presence detection system used in Copenhagen Airport to monitor the flow of passengers-as-bodies. The employed technologies are all biometric in that they measure and identify the *qualities* of bodies, but unlike technologies such as facial recognition, they mainly relate those qualities to spatial parameters and not to delocalized data or IDs. They also relate very differently to the body and the premises for life, as we shall see. It is these differences and their existential and semiotic or sense-making qualities that I would like to dwell on in this chapter.

Borrowing the term *situational awareness*, kindly provided by detection-technology developers, the analysis revolves around issues concerning enskilled senses, categorization, selectivity and the attaching of meaning to amorphous entities, discussing how a presence comes to constitute a threat, as well as the kinds of exchanges that take place around the border fence, framing border work as a cross-fire of mutual scrutiny and surveillance.

Presence detection, Ceuta

Ceuta is a Spanish urban enclave in northern Morocco and is only 19.5 km^2 in area. It has been a European frontier outpost since the early fifteenth century and an important strategic and commercial site for guarding the entrance to the Mediterranean. Changing from Portuguese to Spanish rule in 1668, Ceuta became an essential element in Europe's and later Schengen's external borders with Spain's entry into the EU in 1986 (Gold 2000; Pallister-Wilkins 2017). The old city centre is cut off from the African mainland by an immense fortification lined by a sea-water channel dating back to the Phoenician period and rebuilt by the Spanish in the seventeenth century to avert Moroccan and Portuguese attacks. It now stands as a monumental reminder of the enclave's historically precarious and contested political position.

Besides the amply bridged and mainly symbolic fortification, two borders today separate Ceuta from the rest of the world. To the north, the Strait of Gibraltar, with its strong winds and powerful currents, separates Ceuta from the Spanish mainland and European continent. The strait acts as a 'natural border', being incorporated into the fortifications of Europe and Schengen tactics of deterrence. Migrants attempting to cross the strait from Morocco must overcome the harsh and rapidly changing weather conditions and the immensely strong current, and their failure to do so often goes unacknowledged. These disappearances can be categorized as an act of natural forces for which no political entity or authority can be held responsible. The strait thus plays the same role as the Sonoran Desert along the US–Mexican border, ridding immigration authorities of their responsibility for the deaths, and even sweeping away any evidence of a crime (De León 2015). The open strait is narrow and easily surveyed, being patrolled by both Spanish and Moroccan coast guards. If intercepted by the latter, migrant vessels are returned to Morocco. Crossing the strait by ferry from Ceuta to mainland Spain is another option, as we shall see.

To the south, Ceuta is separated from Morocco and the rest of the African continent by an immense double fence lined with sensors and surveillance technologies. The fencing, initiated in the early 1990s with small, relatively symbolic barbed-wire fences, has today developed into a colossus of steel poles, metal grids, watchtowers, concrete, sewer systems and razor wire (see also Andersson 2016; Pallister-Wilkins 2017; Saddiki 2017). However, despite its impressive aspect and perpetual reinforcement, sustained by ostensibly ever more proficient technologies, the fence is regularly scaled and jumped by huge groups of migrants. As the border guards themselves say, it is only a question of time before it collapses, either physically or symbolically. It is, in fact, the immensity of the fence and its technological 'hardwiring' that, as Ruben Andersson has noted, has triggered the advent of these massive assaults, since 'a critical mass' is now needed to climb it (Andersson 2016).

Both the sea border and the fence are augmented by surveillance technologies. In both border zones, long-range visual and infrared (IR) cameras and radars detect presences at a distance in the upland and sea, whereas different types of haptic sensors are used to detect presences up close. One is the string of haptic sensors built into the fence structure itself, or even skilfully twisted into and hidden in the razor wire. Another type of sensor is the mobile presence detectors that police and border guards take with them when they inspect trucks and containers leaving for the mainland by ferry.

In this particular location, the technologies are thus mainly used to detect unwanted presences, defined as threats. And threatening presences, in this border context, are constituted by two main categories: 'irregular migrants' and stowaways. 'Irregular migrants' are constituted by persons trying to enter Ceuta from Morocco, either by climbing the fence (by far the largest group, primarily sub-Saharan migrants and, to a lesser extent, North African, Middle Eastern and Asian groups)[1] or the much more expensive option of getting a ride by ski jet. 'Stowaways' are those who are hidden within cars passing through the land border crossing, at Tarajal II. They come mostly from the richer sub-Saharan countries, since this is an expensive option or, as we shall see later, in trucks headed for mainland Spain by ferry, an option used mainly by Moroccan citizens. In this chapter, however, I am interested not so much in the actual origins of the different groups and persons who attempt to make it across the border as in how they appear as signs and are read, interpreted and categorized by the border guards in conjunction with presence detection technologies. And, inverting the perspective, I am interested in how particular migrants develop their own technologies and skills in order to read, interpret and circumvent the border and its deterrence modalities.

Beating hearts

'Whenever a truck is getting ready to leave Ceuta, border guards sometimes move in to inspect the loads and containers with their presence-detection gear', Guardia Civil Officer Fernando tells me as we ride through the harbour area on a routine patrol in his green-and-white Guardia Civil SUV. 'The detection gear', he explains, 'is composed of a sensor, similar to a huge microphone, that can be attached to the outside of a container or a truck. The sensor is extremely sensitive to pressure, so it can detect the sound of a pounding human heart'. The system, I learn later, is regulated to filter away 'noise', that is, sounds and other pressure waves that do not correspond to the physiology of a beating human heart and its particular timbre and rhythm, such as those produced by motors, the weather and other animals. Another system enables the officers to detect the sound and the particular rhythms of human breathing. Heartbeat and breathing are, as the technology developers say, 'vital signs . . . which cannot be concealed' (Kurihara and

Watanabe 2011). In other words, the technology relies on the fact that one cannot control – that is, stop – the beating of one's heart or one's breathing without annulling the premises for life itself.

'Hiding as stowaways in trucks going to mainland Spain is a popular method for illegal migrants trying to cross the Strait of Gibraltar', Fernando explains. A carnival is visiting, and in a couple of weeks, the attractions will be dismantled and the trucks will get ready to board the ferry to go back to the mainland. 'The police and Guardia Civil have been carrying out a border guarding operation, Operación Fin de Feria, for some years at the end of the carnival. Last year they had pretty good results', Fernando says,

> catching several illegal migrants with their presence-detection gear. In the back of a truck, the sensors had identified the signs of a beating heart – and then another – and another. Using their dogs to smell the exact locations of the intruders, the officers crawled into the truck and extracted several adult men who were hiding, twisted around the elements of the carnival attraction like contortionists,

as Fernando pictures it. The owners of the carnival attractions were suspected of trafficking but were later released, claiming they had no idea they were carrying stowaways and that the migrants must have hidden in the trucks during the night. The prospective migrants were all Moroccans who had legally entered Ceuta on a day pass that did not allow them to travel on to mainland Spain. Their hearts and their breathing had betrayed them.

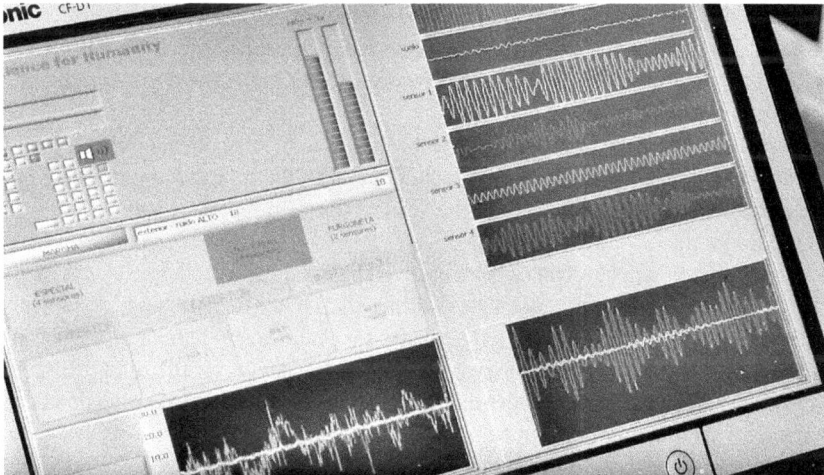

Figure 4.1 Screen of heartbeat detector
Source: Photo by the author.

The presence sensor functions like a gigantic ear, glued to the surface of the truck. The movement of its membrane – picking up vibrations, filtering them and transforming them into signs of life – simulates the movement of a listening human tympanum, simply bigger and more fine-tuned. The agent using the sensor has a screen and a loudspeaker at his disposal, so he can see the vibrations from several human sources, transformed into visual signals, and to hear them if he wants to. Each of the different hearts shows up as a graph on the screen, showing every heartbeat as a peak. From the analysed signals, defined as human or as interferences, the agent directs the search for the stowaways' bodies and their incriminating hearts.

So, like fingerprint scanners (see Part III) or the facial-recognition technologies discussed in Chapter 3, the heartbeat sensors work on the basis of iconic sign relations, searching for likenesses between an object – the rhythm and timbre of a beating heart in a stowaway body – and the registered and stored templates of known human heartbeats. That is how the filter recognizes a human presence. But contrary to fingerprint and facial-recognition technologies, the presence detectors can hardly be circumvented, and the likenesses cannot be dodged (see Chapter 3). One cannot make one's human heart sound like a dog's heart or like a motor. There is indeed something of Giorgio Agamben's well-trodden 'bare life' at play here – a bare human heartbeat, unalterable, uncontrollable – or of the unescapable belonging to a single category, the ontological collective that is the human species subordinating all other categorizations, all other profiles, to what is culturally constructed (Rapport 2017: 6). Only the human beating heart is the matter.

The sensor defines an intrusion by coupling it to a particular place, always a place not intended for human presence. It is indeed the hiddenness, the 'stowed-awayness', that defines the presence as a threat, as a body that requires interception, not the identification of a particular individual. In semiotic terms, we could say that it is the direct, indexical connection between the sensor and the beating heart that provides the sign, the evidence of a presence, and the place that defines the presence as illegal and a threat.

Fernando shows me photos of migrants whom he and his colleagues have found hidden in car seats or bumpers or wrapped around motors. Every photo seems to attest, in his eyes, the efficiency of the sensors but also the ruses and desperation of the migrants.

The heartbeat detector can only be used in confined spaces, such as trucks, containers and cars, where the resonances of the sounds are not jumbled by interference. Its ruthless accuracy, rendering almost superfluous any human intervention, any interpretation of the signals by a border guard, cannot, for the time being, be applied to open spaces and longer distances. But other types of presence detector, haptic, visual and thermal, can be used in the open, as we shall see next.

Fence haptics and other techno-sensory engagements

Along the border separating Ceuta from Morocco runs the 8.5-km-long, 6-metre-high double fence armed with razor wire and a string of watchtowers, floodlights,

Figure 4.2 Patrolling the border fence with the Guardia Civil, Ceuta, 2017
Source: Photo by the author.

daylight and IR surveillance cameras and haptic sensors, complemented by mobile IR cameras mounted on the SUVs that regularly patrol the border at night. The fence cuts through the undulating landscape like a sawtoothed knife, running through deserted gorges and valleys and between houses and gardens where animals stray and children play on each side, stretching into the sea at each extremity, over beaches where tourists sunbathe and swim. All along its sinuous body, the string of surveillance devices clearly indicates that at the end of the connection someone or something is alert and watching.

In the COS – Centro Operativo de Servicios – above the Tarajal border crossing, Isabel, a senior Guardia Civil officer, is surveying her screens. On her desk, she has a row of smaller screens, and on the wall in front of her, four large screens each depict four images from the more than 60 daylight cameras along the border fence and its sewers. The images automatically shift to new cameras every 15 seconds or so. A mechanical voice issues an alert whenever the haptic sensors detect movement along the fence. On one of the screens we can see two men repairing a big hole in the fence, made by a group of migrants attempting to cross into Ceuta early in the morning. They only made it through the first fence, being intercepted at the second fence by Isabel's colleagues and returned to Morocco. Because of the repairmen's work on the fence, the voice repeatedly alerts Isabel of an unknown presence in section G/camera 19, and the corresponding camera image stays on. It is distracting, and Isabel would like to turn the alert system off. 'But I can't', she explains, 'in case someone tries something somewhere else'. She looks at the screens as the images from the 60-odd cameras continually shift and move our attention to new sites along the fence. I cannot make out where they are pointing, but Isabel knows exactly where she is. She knows the landscape by heart, moving through it via the surveillance cameras, she says.

Isabel can also turn the cameras around and scan the landscape with a joystick. They do not scan automatically – that would be a nuisance, she says. The cameras do not know where to look or what to target, and it would be a mess if the images were moving all the time. They cannot distinguish important from unimportant, sense from nonsense.

The alerts from the haptic sensors are often triggered by stray donkeys, dogs, large birds, deer and horses passing by the fence, and Isabel just saw something that looked like an ibis, only smaller. She has become quite skilled in animal biology, she says, sitting here for long hours, distinguishing the living from the inert, humans from non-humans.

A few of the images do not shift. The cameras are pointed at the big sewer holes in the fence's foundations that prevent the whole construction from collapsing during heavy rains. The sewers are the fragile points in the fence, she explains, because they are so big. A person could easily walk through them upright.

At another workstation, two screens show images from IR cameras that are used at night to detect body heat. On one of them, humidity completely covers the lens. But when night falls, patrols will move along the fence with mobile IR cameras mounted on tall poles. The fixed cameras, as well as those mounted on the patrol cars, are pointed into Moroccan territory. 'When a group of people arrives at the fence, it's already too late for us to react', Isabel explains. 'When they touch the fence, they're already climbing it'.

So the cameras, which point into Moroccan territory day and night, and the images they display on Isabel's and her colleagues' big screens, seem to be the pivotal tool. But the cameras cannot see – they need human operators to make the distinctions between humans and donkeys, between local residents and migrants, between the anodyne and the threat.

Isabel's colleague talks about a border guard who worked in the COS some years back. He mainly took night shifts, surveying the screens of the infrared cameras. After a while in the service he could tell the difference between a Maghrebi and a sub-Saharan. I ask them how he could see that from something as indistinct as an IR image, just a white figure against a dark background. 'He could see it in the shape, the posture, the way of moving and walking, and the clothing'. This was not something he had learned in a training program but was of the order of tacit knowledge, even if his colleagues could recognize and describe it. 'He had a special gift . . .', Isabel's colleague concludes.

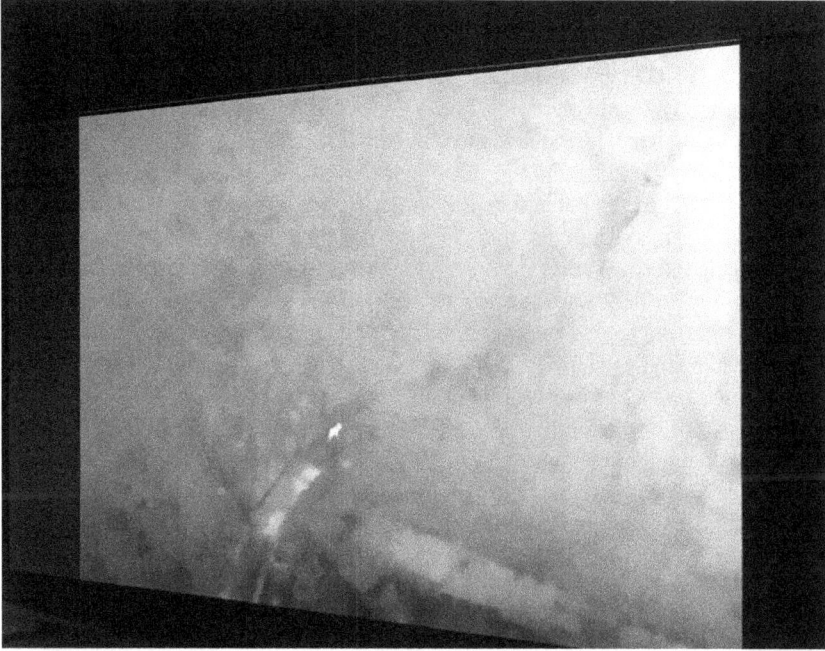

Figure 4.3 Developing zoological skills with the IR surveillance camera
Source: Photo by the author.

The COS is the central nervous system of an infrastructural installation con-sisting of the fence and its sewers, watchtowers and hinterland, hooked up to the COS through a sensory network that can feel the presence of bodies, sense their body heat at a distance and see them as they move and close in on the structure. At the end of the electric nervous system, in front of the screens, human percep-tions and interpretations take over the sensory work, attributing meaning to these technical sensations. The figure of the cyborg comes to mind, for the operators in front of their screens form an integral part of this sensory system, manipulating the joystick, switching between images, picking up the alerts of the haptic sen-sors, scrutinizing the images for details, interpreting all these inputs and, on some occasions, giving orders to mobile units to move out to spots of intrusion.

A cross-fire of surveillance and technological enskilment

The officers in the COS and on patrol along the fence use high-tech presence-detection devices to survey the border zone up close and at a distance, but they are not the only ones on guard. Indeed, as had happened this morning, large groups of sub-Saharan migrants regularly attempt to cross the fence, sometimes success-fully. They have been lying low in the surrounding Moroccan mountains scruti-nizing the fence, the technologies and the routines of the border guards, becoming techno-specialists in their own right. While waiting to work with the Guardia Civil officers in the COS and on patrol, I have been discussing the border technologies with a group of men who managed to cross the fence in February 2017. They seem to know the border zone and the fence, with all its devices, capabilities and fail-ings, as well as or even better than the border guards, for the border has been their focal point for many months, their main preoccupation 24/7. Most of them had tried to scale the fence several times before actually succeeding one early Febru-ary morning. So they know the fence from close, personal acquaintance; they have been cut and snared by it, trapped, deterred and severely wounded; and the signs on their bodies show it. They have an intimate and wretchedly embodied, even haptic knowledge of its materiality.

'Frappe 358', they call their last and final attempt: 'Hit 358'. Like all the other 'hits', it is named after the number of people who managed to cross. And they have all defeated it and can talk about their technological knowledge and skills with pride. One of them, François, gives me details about the fence, the number and type of cameras installed, the routines of the patrols and the watchtower man-ning on both sides of the border. Over the weeks, he also talks to me about how the migrants organized themselves in the mountains and the many months they spent surveying the fence. I come to learn about the fights among the migrants; the changes in leadership; the regular coups d'état where the leaders and most of their 'government', as they called it, were overthrown; the conflicts and alliances between national groups; the fate of snitches; and the logistics and preparation of the 'hits'. Without him telling me, I sense he is part of the leadership, as indi-cated by the way the others still relate to him. And he confirms that, although the organigram is no longer officially on display because those in leading positions would be charged as traffickers and held responsible for the damage to the fence and the border guards, the internal hierarchy will endure even as time goes by and the surveillance will continue from within Ceuta. He has clearly played a leading role in the organization and tells me about their intelligence service, *their* border work. They surveyed the border with binoculars and even, at some point, a night-vision monocular, and they were in constant touch via mobile phones.[2] They could track the movements of patrols, detect their routines, count their numbers and keep track of their schedules, nights, days, weekdays, weekends and holidays. And over the months, they would get an idea of the Guardia Civil's technologies, how they used them and when they failed to work properly, that is, in heavy rain, fog, strong wind, smoke or strong sunlight. They singled out the IR cameras that were not working and noted when they were undergoing maintenance. They figured out how topography and weather played a part, when it was to their advantage, the fragile areas and when and where to 'hit' by sur-prise. And one early morning, they moved in on the border in one of those fragile zones where the technologies were failing and patrol access was difficult. They lay low on a mountain slope for 24 hours, covered only by low bushes while the

Moroccan forces unknowingly changed shifts in the nearby watchtower. And when the time came, all 358 men rose out of the bushes, stormed the fence and managed to cross it and run all the way to CETI (Centro de Estancia Temporal Inmigrantes) before anyone noticed.

They continue to survey the fence from the other side. CETI, where they are lodged while waiting for their expulsion order – paradoxically executed by giving them a laissez-passer and a ferry ticket to the Spanish mainland – was only designed to hold 512 migrants but was accommodating 1100 at the time of my fieldwork. They know that when a new group manages to break through the fence an equal number of CETI residents will be sent across the strait.[3] Without saying much more, I understand that they also engage in digital communications and intelligence networks across the fence and that those exchanges also play their part in the border work, just like the agreements between Spanish and Moroccan police and border guard agencies.

After the 'hit' in the morning, the Spanish minister of the interior announces that the border in Ceuta will be made *más inteligente*, that is, 'smarter'. Drones will be sent in to patrol the border, also providing surveillance of the most inaccessible areas where most of the 'hits' take place. Fernando and his colleagues roll their eyes. They know very well that there are 'dark spots' that are difficult to survey and patrol, that the migrants have the upper hand because they can choose the time and place and take the border guards by surprise and that they will always come and cannot be stopped. And they also know from first-hand experience that technology is not the answer. They acknowledge that the migrants have good reasons for wanting to move into Europe and that the solutions are rather of a political and economic order, as well as being on a much higher level, out of their hands. They seem quite disillusioned. As for the drones, 'Who is going to fly them? And who is even going to look at the images and interpret them?!' There are already only very few people on duty every night to patrol the fence and survey the screens in the COS and, Fernando adds, the things won't even be able to fly in the narrow gorges of the border: 'This is Gibraltar; the winds are far too strong, they will crash into the mountain walls within minutes!' Some months later, the minister rolls back his proposal to purchase drones for border control in Ceuta.

Reading signs, detecting threats

In all these border control settings, visual, aural and haptic presence-detection devices register and measure bodily qualities and detect bodily presences. Coupled with human interpretative skills, and sometimes even special 'gifts', it becomes possible to define their kind. The technologies range from complicated seismic devices with digital filtering and IR detection of body heat to tactical communications and direct visual inspection and analysis. In all cases, the presences detected make sense as threats because of their spatial positions, the locations in which they are situated, whether hearts pounding in a container or bodies moving in on a fence. The categorizations and profiles are made on the basis of assumptions vested in these localized bodies about their past and future moves, not on their identities and on previously registered data. On the screens, the bodies show up as simple graphs, dots or vague outlines, signs that need to be interpreted and given content for action to be taken. Is it a dog, an ibis, a human? Is it a threat? For in the end, border control is all about reading signs, whether the interpretation is done by an algorithm, a device, a human or, as most often, a combination of these.

Presence detection, Copenhagen Airport: the surveyors surveyed

Presence detection is also used in other settings where the vague figures take on other meanings because they are inscribed in other spatial, political and economic configurations. To both broaden and sharpen the analysis of the interpretive work of presence detection and to understand how vague contours take on meaning depending on their contextual organizational and political configurations, a final example is briefly presented. In this example, the dots also constitute bodies in movement near a border, but the *objects* of control are not the bodies themselves but the system *impeding* their movement. In this case, the object of control is itself a border police unit, creating queues and obstructing flows in an international airport. The example serves to turn around the perspective of figuration and control once again in order to further our understanding of the scope, functioning and semiotic qualities of these types of biometric surveillance technologies and of the particular kinds of 'situational awareness' and threats with which they work.

In the main working space of the Border Police in Copenhagen Airport, a big screen on the wall shows a black-and-white still-image of the floor in the

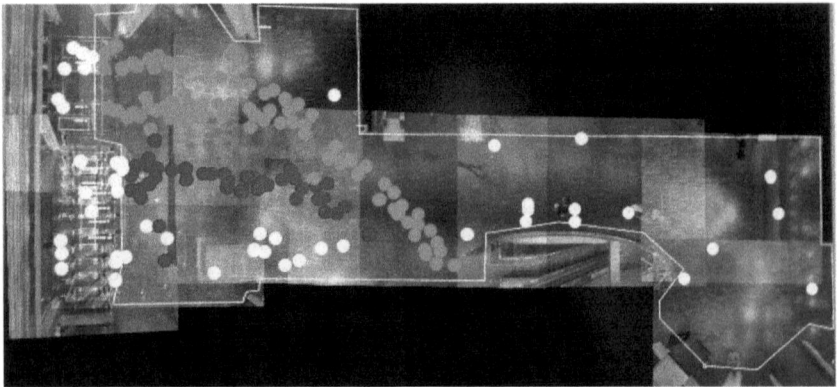

Figure 4.4 Flow detection, Copenhagen Airport
Source: Photo by the author.

border area. The inert image becomes animated when a person moves into the zone. '3D people-tracking sensors' installed in the ceiling of the border zone detect the human presence, and a small white dot starts moving across the image on the screen. If the person queues up at the manual passport control, the dot turns bright red. If the person slows down in front of the Automated Border Control, the dot turns blue. If all goes well, all the white dots remain white, meaning they are in movement, only momentarily halted by the border control.

At present, the many red dots have turned into a big glowing red blob on the screen, indicating that too many people are queuing in the passport control instead of reaching their flights or lingering in the shopping areas. But in the police office, no one looks at the screen except for the anthropologist, who is fascinated by the aesthetics of the imagery, the small moving dots that change colour and the fact that the border guards are themselves being monitored and controlled. The border guards, on the other hand, do not look at the screen. They know perfectly well when people are crowding: they can hear it through the thin walls, sense the bustle and see it from their service centre, the 'aquarium' that overlooks the border zone.

However, although the police officers no longer notice the screen, the red and blue dots are certainly seen, registered and audited in the offices of CPH Airport A/S, the private company running the airport. And they will use the figures in their next negotiations with police management and Frontex about increasing flow and reducing retention on the Schengen border by lowering the settings of the threshold of resemblance in the ABC (see Møhl 2018).

The bodies registered by the tracking devices in the airport are anonymous, like the migrants trying to move in on the fence or across the strait. Their identities are not important. They are, until proved otherwise, devoid of anything other than potentiality, posing different kinds of possibilities and threats. But threats come in many forms, and here the bodies, rising not out of the bushes but in the lack of flow, directly threaten commercial interests. The border police wish to fulfil their obligation to protect the border against what are defined as threats to national and European interests, whereas CPH Airport A/S wishes to have more active consumers in the shopping areas, even at the cost of lowering vigilance. These two threat images and their respective agendas clash. The questions, however, are whether the motivations for guarding the borders of Ceuta and Europe are, in fact, so different and whether economic agendas do not also play just as decisive a role in the overall policies of erecting border infrastructures and investing in ever more proficient technologies. This is suggested by several analyses of borders and migration deterrence as sites of industry, economic investment and spectacle (e.g. Andersson 2016; Andersson 2014; De Genova 2013; Gammeltoft-Hansen and Sørensen 2013). In these different presence-detection systems, travellers, migrants and border guards all become *objects* of policy (Feldman 2013): different types of policies, different agendas, yet all are part of the flow and management of the border world.

Situational awareness and the nature of blankness

This chapter has raised several questions concerning the semiotic and political processes involved in characterizing seemingly void figures and the nature of a presence. How does one attribute meaning to a white, anonymous dot or figure? How does one learn to see and read them, vest them with intentions and see them as threats? What are the categorizations and selectivities involved, when do they operate on an individual basis and when are they collectively enskilled and formatted?

First, the presences that appear as heartbeat signals and white or red figures on screens are simply biological presences, devoid of personal identity, devoid of any individual characteristics. In terms of biometric identification, they are 'data blanks', 'whats', not 'whos' (Feldman 2013; Rapport 2013). In Michel Serres's terms, we could call them 'white multiplicity', 'blank figures' that could be vested with any identity, any scenario about the past and the future (Hetherington and Lee 2000; Svendsen 2011). The 'blank figure' seems to relate quite well to these anonymous dots on the screen, stripping the moving bodies of all predetermination. Every dot is possibly *an anyone*, a 'cosmopolitan anyone' (Rapport 2017) – but in an ideal world, then, one without borders. For the anonymous and unidentified dots are mobile and vectorized: they have a direction and a presumable goal in the eyes of the border guards. In that sense, they already have a story.

On the nature of blankness and writing about the painter's white canvas, Deleuze holds that it is anything but blank.

> It is a mistake to think that the painter works on a white surface. . . . [E]very-thing he has in his head or around him is already in the canvas, more or less virtually, more or less actually, before he begins his work.
>
> (Deleuze 2003: 71)

In the same manner, he says, the 'blank page' is already full before we start writing. In other words, the slate, canvas or figure is never blank, *rasa*; in our perception it is already saturated with stories and images, possible pasts and futures, from a variety of sources. The painting is made not by the painter in an artistic void, but in an interplay between the artist's body, senses and intentionality, and complex, shifting political, ideological, aesthetic and economic forces. Deleuze uses the painter's work and white canvas as a means to think about society and politics, as well about sensations and meaning. Following this allegory, the border guards' sensory work takes place within particular political and technological settings. More concretely, as we saw in Chapter 3, their senses are enskilled and formatted not only in isolation but also in a particular community of practice (Grasseni 2007b, 2011) – and, we might add, in an interplay with the technologies that form and are formed by human perceptions and agendas. The white dots on Isabel's screen, then, do not simply constitute Michel Serres's blank figures but are also already vested with meaning, with significance and presumed goals, because they are localized and vectorized. For example, unidentified threats, unspecified enemies. A donkey, a Moroccan and a sub-Saharan are all living bodies, all white figures, but with very different political and practical meanings and immediate consequences for border control.

The unidentified presences that move about Copenhagen Airport and queue in front of the border control, forming big, colourful blobs, are not blank either, even if they are merchandized as 'anonymous' (XOVIS AG 2019). To the XOVIS system and its tentacles, they have the identity of potential consumer bodies. And when they are transformed into blobs, they constitute dense signs of occlusion and lack of flow. They are figures played out in the negotiations between private and public interests: between the airport's desire to create a seamless experience of pleasure and consumption and the authorities' need to control identities. And, like the Threat Image Projection images described in Chapter 3, they are converted into numbers and used for negotiations when the border control agency and the individual border guards' productivity are audited at the end of the month. Analysing their particular and somewhat simplistic mercantile and auditing semantics clearly points to the semantic density of seemingly blank figures that could be vested with any story but that usually have a quite clear – here, commercial – story, written directly into both the technologies and the policies surrounding their installation.

Zooming in on the very minute daily interactions between technologies and bodies, both surveyed and surveying, enables us to understand more about the sensory activities that lie at the basis of every human activity, including border work, the fine balance between technological detection and human interpretation, and how different bodies are constituted as more than neutral objects and come to stand as threats. It also puts into perspective some illusions we might have about the efficiencies of the technologies in use and sheds light on the discrepancy between political narratives about the character of the threats and the infrastructural means to deter them, on the one hand, and the sensations and experiences of those who engage in border work on a daily basis – who meet the 'threats' in the form of real human beings and go about their tasks, often without much illusion about their utility – on the other. Inverting the perspective paints a larger picture that includes both internal ambiguities and mutual awarenesses and draws a more nuanced picture of what separates and what unites people on both sides of the border, both in their daily tactics and in their technological-sensory work.

Notes

1 During my fieldwork in 2017, an average of 1100 individuals were staying in the Centro de Estancia Temporal Inmigrantes, waiting to have their cases processed; a total of 2244 'irregular migrants' entered Ceuta that year (El Faro de Ceuta 2018). The great majority, approximately 70%, were sub-Saharan men who had entered by jumping the fence and who were not requesting asylum, plus just a handful of women who had passed the border as stowaways in cars or on jet skis. The rest were mostly Algerian, Syrian and Sri Lankan men.

2 Another young man describes how, before jumping the wall, they burned all their cell phones to avoid tracing and police intelligence work. In a similar vein, Ruben Andersson

describes the tactics of migrants using GPS and compasses and throwing them over-board before arriving in the Canaries so not to be seen as captains and traffickers (Andersson 2010: 37).
3 Once arrived, they are placed in temporary centres for a maximum of three months, after which they are required to return to their countries of origin. Complying with the expulsion order was, however, not part of the plan of any of the persons I discussed this with.

Epilogue

Knowns, unknowns and the ubiquity of uncertainty

'You can look like someone else all your life but you can only hold your breath for so long', a border guard once said to me when we were discussing the efficiency of ID control and presence detection. In a sense, he was summing up the fundamental difference between facial recognition technologies like the Automated Border Control described in Chapter 3 and the quite simple presence detection technologies discussed in Chapter 4.

Facial recognition involves complex and very high-tech operations and calculations, but it can be circumvented with sufficient mimical mastery. Presence detection simply amplifies human senses; it cannot be circumvented, or only with difficulty – at least not within the limits of living life and only by undermining its sensitivity by 'hiding' behind other sound sources that would interfere with its hearing. So facial recognition might be promoted by both lobbyists and policymakers as an efficient and foolproof solution to border control, surpassing both human capacities and their biases. In practice, however, its authority is located in the myth of an irrefutable one-to-one relation between a human identity and a tiny ID photo, while its efficiency is questioned by police and border guards who are constantly seeking to test its capacities, infallibility and defects. 'Can it still tell the difference between me and the ID photo of a chubby someone else?' 'Has someone tinkered with the recognition threshold overnight to comply with commercial interests?' 'Are my sensory skills, my intuition and my foresight – in short, my profession – being undermined by economic interests and decision-making?'

Comparing manual and automated border control, the ABC displaces the ability to detect from a human being to an algorithm and its capacity to recognize. That is how it is presented by developers and how it is experienced by travellers and border guards. This changes the interaction and thus also the relations established between controlled and controlling force. Passing through the biometric eGate, as a traveller, you cannot look the controlling agent 'in the eye', at least not a facially embedded eye. The eye of the camera is nebulous, its neural networks are connected to unknown agencies and its seeing comes across as opaque to the seen; the controlling function has been prolonged into the past, and it has been extracted from the individual border guard and delocalized from the border itself, taking 'place' in and operated by distant servers and organizations.

Common to all border control technologies and practices, however, is the permanence of the great unknown: How many persons actually managed to outwit the technologies or our vigilance and cross the border? There are no figures for that great unknown, only guesses, which, by definition, are undocumented and unqualified. The spectre of uncertainty is particularly discernible in the airports of Copenhagen and Gibraltar, on the land border between Gibraltar and Spain, at the harbour in Ceuta and at the border crossing point at Tarajal II. In all these sites, it is impossible to know how many people pass the border 'under the radar'.

But there is one exception, one setting where the unknown has yielded to an unmistakable *certainty*, a certainty that is produced by the particular geography in conjunction with Spanish immigration policies. The big known – which in itself becomes a key site for the confirmation of imagined threats – materializes on the land border between Ceuta and Morocco. Here, it is conspicuously *known* quite exactly how many people have managed to cross the border by scaling the fence. And this is because, once they have crossed the border, the passers become known to the authorities, either because they are directly intercepted by the Guardia Civil or because they make their own way to the Centro de Estancia Temporal Inmigrantes, like the 358 who made it across early one February morning in 2017. Only a few of these people will ask for international protection. The vast majority will be given an expulsion order and then wait till they are granted passage to mainland Spain. Once they are on the mainland – and maybe after some months' stay at a migrant reception centre, after which they are expected to return to their home countries – they will go underground and attempt to find housing and work, usually ending up in the extensive circuits of undocumented and exploited migrant workers. Only *then* do they join the ranks of the unknown and unregistered, but they no longer pose a direct threat to the border and the border guards, as they have entered another domain of European migration policies.

The newspapers in Ceuta punctually register the successive border fence crossings, often with spectacular videos and photos of huge masses of people perched on the tops of the fence, ensnared in the razor-wire, or running victoriously through the streets of Ceuta yelling victory – '*Booza!*' Occasionally, the border authorities release particularly spectacular images and videos to support their claims that they are violently attacked on a regular basis and in need of general reinforcement. On August 10, 2017, a few weeks after I had left Ceuta, a group of two hundred migrants managed to enter Ceuta by forcing their way through the border crossing at Tarajal II, to which they resorted after being deterred from climbing the fence.

According to the media, the border at that point was manned by police officers, who are normally engaged only in document control, not by Guardia Civil officers (Echarri 2017). A surveillance camera inside the border crossing filmed everything, and soon after the incident, the border authorities released the video (El Faro de Ceuta 2017). On it we see casual early morning routines when all of a sudden, at 4.50 a.m., a person jumps the gate and runs through the border passage

Figure II.2 Tarajal II border passage, 7.08.2017 (excerpt from surveillance video, El Faro de Ceuta 2017)

unhindered, followed by many others, while the police officers stand more or less passively by. One officer tries to hit the migrants with a baton, while another tries to kick them. The latter stumbles in his attempts and breaks his lower leg. The fractured and twisted leg can be seen very clearly on the video – which is probably why it was released. And because of the intensity of the direct image and the fracture itself, which from a superficial viewing could be attributed to the passing migrants, the video goes viral and quickly makes it not only on to the national but also to international headlines and media. This leads the interior minister to announce that even more efficient measures will be taken to prevent border crossings, including the installation of more intelligent borders armed with more high-tech solutions. As such, not only through its impressive content but also because of the routes it takes and the ripples it creates in the European border world, the video and its repercussions add to the intensity of the border spectacle and provide an impetus for even more investments in border control technologies of various sorts. And, in 2018, it is decided to rid the fences at Ceuta and Melilla of the razor wire that does so much visible bodily harm and to replace it with ever more efficient technologies. In April 2019, as I write this, the pope has just visited Ceuta and, a reporter writes, wept when he saw the razor wire. Spectacle, symbolics, blood and tears, all too conspicuous, all too human – whereas the processes that take place 'inside' the technologies seem to be lifted from the arena of morality, human intervention and judgement, being liberated even from the domain of responsibility and politics.

Figure II.3 Open border, Gibraltar/Spain. Border spectacle, symbolics and doing away
       with uncertainty

Source: Photo by the author.

Border spectacle, symbolics and doing away with uncertainty: On the heavily symbolic border between Gibraltar and Spain, the Spanish border authorities have installed 26 ABC eGates at tremendous cost. However, they are often out of order for days on end. Instead, a sign announces that the floor is wet and an open door lets travellers pass the border unhindered. That is, one might say, one way of doing away with uncertainty and the unknown: opening and thus eliminating the border altogether.

Part III
En route

Anja Simonsen

Introducing the site

Migration is and always has been 'an integral aspect of the life trajectories' of human beings across the globe (Olwig and Sørensen 2002: 10). We have moved and are still moving, for example, as 'explorers, colonizers, traders, tourists, refugees and labour migrants' (Rytter and Olwig 2011: 9). Hence, despite the fact that the media and politicians today usually portray migrants as poor people from the Global South, a very diverse group of people located all over the world 'live in mobility' (Schrooten et al. 2016). There are many different reasons why people decide to move: looking for new economic opportunities; acquiring further education; escaping from poverty, war, conflicts and climatic problems; or just pursuing a nomadic lifestyle (see e.g. Hansen 2006; Dalgas 2015; Schrooten et al. 2016; Kleist 2016). As a Somali writer explained to me during an interview: 'It is the nature of human beings to leave a place where there is no life towards a place where there is a life [. . .]'.

The world in which we move has changed dramatically during the past 20 years. Contrasting perceptions have been put forward about how integrated 'mobile livelihoods' (Olwig and Sørensen 2002) are, and should be, in livelihood strategies across the globe. While many people consider freedom of mobility to be a natural part of life, as the Somali writer's earlier comment shows, recent migration management initiatives in the form of increased biometric border controls seek to limit free movement to those who are in possession of legally valid documents. As a consequence, migration and mobility, in general, are experienced in very different ways by different travellers. For those who have the requisite documents, migration is fast and seamless. For others, such as those who are drawn towards Europe because of civil war and continued conflicts,[1] biometric border control has made mobility increasingly difficult. Indeed, the undocumented mobile body will not only encounter increasing biometric border controls, whether at territorial borders dividing countries in Africa or at ports of entry from the African to the European continent; it constitutes the material for biometric border control itself in the form of, for example, fingerprints that can be registered and identified or a beating heart that can be detected.

Part III – *En Route* explores the contemporary migration practices of Somali women and men as they move towards, and within, Europe in search of more secure livelihoods while confronting new European initiatives on migration management and border control. More specifically, it discusses the implementation of the biometric registration of fingerprints at particular European border sites – 'hotspots' – and shows how such biometric practices have turned the very bodies of migrants[2] *en route* into a space of border control.

This form of border control has not always existed. As the Somali writer interviewed earlier explained:

> Somalis used to migrate in the past when it was not illegal because there were no borders and no political regulations. The land was the land for human beings, not for Kenyans, not for Somalis; you could go everywhere.

The phrase 'land for human beings' refers to the time before the colonial powers divided Somalia into five parts without taking into account centuries of 'nomadic travel, spiritual interaction, and trade' (Weitzberg 2017: 2). Before colonialization, many Somalis identified as members not of nation states but 'of Islamic, nomadic, and lineage communities that spanned Northeast Africa and Arabia' (Weitzberg 2017: 4), which allowed them to live 'mobile livelihoods' (cf. Olwig and Sørensen 2002).

Part III sheds light on how in Italy the implementation of borders, in this case biometric borders, has meant that the Somali migrant community located there must make new choices and adopt new migration practices. Adapting to such obstacles by 'circumventing, redrawing and rethinking' borders is nothing new, however (Weitzberg 2017: 2). Chapter 6 shows how Somali migrants' current practices of sharing information constitute a tactical management of border-like experiences of migration and that these tactics resemble past approaches to obstacles in the form of borders. This holds true whether the border is in the form of a harsh climate, such as a hot, deserted desert or a violent sea that is very difficult to cross, clan conflicts and civil war or territorial borders between nation states guarded by border police and dogs. In his ethnography of the northern pastoral Somalis, I.M. Lewis describes how their social and economic lifestyles centred on moving from pasture to pasture (Lewis [1965]2002: 1–12) and how a greeting such as *ma nabad baa* ('Is there peace?') would be called out when meeting a stranger as a request for information on what might lie ahead (Lewis 1993).

Migration out of Somalia towards Europe began during colonialism when Somalis started working as soldiers and sailors for the British. This eventually led to the establishment of Somali communities in Cardiff, London and Marseilles (Kleist 2004: 1–4; Hansen 2006: 66; Diiriye et al. 2015). Others migrated to Italy as students or military trainees during the 1950s (Fagioli-Ndlovu 2015: 7). It was not until 1991, however, when civil war broke out as a result of the collapse and overthrow of Mohamed Siyad Barré's regime, that the number of Somalis fleeing their country increased dramatically, and many set out to settle in Europe and North America. The two major cities of Somaliland,[3] Burco and Hargaysa, were bombed between 1987 and 1989 when Siyad Barré was still in power. This resulted in hundreds of thousands of people fleeing, more than 50,000 deaths, and Hargaysa being rendered a ghost town (Hansen 2006: 20, 56). In the rest of Somalia, ethnic cleansing, violent conflicts, famine and starvation broke out (Lewis [1965]2002: 262–265), the capital Mogadishu being 'the centre of waves of destruction and terror' (ibid.: 264). Out of a population of approximately 15 million people in greater Somalia, about 1 million Somalis have become part of the mixed migration flows within the Horn of Africa region living as internally displaced persons or in refugee camps. Another 1.5 million Somalis have resettled as refugees in other countries in North America and Europe as part of the diaspora. Another 1.1 million Somalis live as internally displaced persons within Somalia (HRW 2018). Conflicts within Somalia between the militant group al-Shabaab (the Youth)[4] and the government continue to rage, as manifested in the internal and external displacement of large parts of the population. In October 2017, 358 people died as a result of a bomb attack in the centre of Mogadishu, adding to the 1228 civilians in Somalia who were reported to have been killed between January and September 2017 (HRW 2018).

The majority of the Somalis who are located in Europe have fled from the previously mentioned wars and conflicts. The remaining have fled from what today is known internationally as the peaceful part of Somalia: Somaliland. Despite the fact that Somaliland has adopted its own constitution and political system after its independence from the rest of Somalia in 1991 (Hansen 2006: 9), young Somalis continue to flee. The Somalis I interviewed during fieldwork described the reasons for fleeing as being related to a lack of infrastructure, education, gainful employment, marriage opportunities and widespread poverty. Facing such obstacles created feelings of hopelessness and disparity that eventually led them to migrate. Somalis refer to the contemporary way of migration, which is practised mostly by the young Somali generation, as '*tahriib*'. *Tahriib* constitutes a search for a sustainable life far from a context that is generally experienced as insecure. It involves venturing out on an uncertain journey as migrants without the necessary documents to cross contemporary borders (Simonsen 2018).

These young people are motivated not only by a desire to improve their own life but also by an ambition to help and care for family left behind. As I show in Chapter 5, they often end up becoming dependent on care themselves, however, as they fail to reach their migration destinations due to biometric border control implemented by the EU. Migrants' routes, and the lives they enable and disable, are thus entangled in a web of complex and conflictual practices of care.

The biometric body border

Somali women and men face two forms of borders as they attempt to cross into and move across the EU. The first type of border they are confronted with is territorial. When they leave Somalia, they must pass an endless number of such borders that demarcate where one nation state ends and another begins. These borders are often demarcated by fences and guarded by border patrols. Before the implementation of fingerprint registration, iris scans and presence detection equipment, as discussed in Part II, migrants were often able to negotiate their movement across such borders by using printed ID documents of various sorts.

The ways of negotiation depended on which types of ID documents they possessed. If the documents were authentic, they were borrowed, often for a fee, from a Somali located in Europe. This meant that the migrants had to obtain a document from a person who resembled them (see also Chapter 3). It also meant copying the look of an immigrant living in Europe and learning basic conversation skills in the language that this ID document was issued in (Simonsen 2017). The ID document acted as a form of a mask that had to be put on when crossing territorial borders (Kelly 2006: 100). If the ID documents were forged, the risk was higher for the migrants. In such situations, it was important that the forged document closely resembled the latest versions of the document, for example, a passport, as the migrants would otherwise risk being caught when seeking to travel farther. It was well known within the migrant communities where I circulated that being caught with a fake passport was punished more harshly than travelling on an authentic but borrowed one (Simonsen 2017). It has become more difficult to use such documents due to the implementation of biometry into ID documents, as is illustrated in the following two chapters (see also the work of Coutin 2003; Kelly 2006; Reeves 2013).

The second type of border that the Somalis *en route* encounter is an integral part of themselves: their body. The implementation of biometric technology, such as fingerprint registration for identification purposes and asylum processes, has turned the bodies of migrants into borders of their own. Wherever women and men *en route* move, they are bringing their own individually embodied border that, if live scanned, will share the information of their past registrations and thus whereabouts with the authorities. Some have already encountered the body as a border in refugee camps in the Horn of Africa when seeking assistance from global organizations such as UNHCR (see Chapter 5). Others are not confronted with this type of biometric border landscape until they reach Europe. The majority of Somalis who reach Europe arrive by boats that have crossed the Mediterranean, and they encounter physical European borders in Southern European countries such as Italy and Greece. Since they do not have the required documents, they can only enter Europe legally by seeking asylum, which involves having their fingerprints registered. Their fingerprints are stored in EURODAC (see the Introduction) and document where and when they entered Europe and sought asylum. The authorities will use the information stored in EURODAC in the subsequent asylum-seeking procedure, following the Dublin Regulations (see the Introduction).

The management of European borders changed in 2015 when the number of migrants arriving in Europe peaked. Before 2015, many of the migrants who entered Southern Europe, arrived outside official border control posts and therefore avoided fingerprinting and were not registered as asylum-seekers in their country of entry (European Commission 2015a). In Italy, it was the responsibility of the migrants to present themselves to the Italian authorities within eight days, in accordance with the Italian immigration legislation, to be eligible to initiate an asylum request (ASGI 2019). This is still the case today. In April 2015, however, the European Commission implemented the so-called hotspot approach at five locations in Italy and Greece (European Union Agency for Fundamental Rights

(FRA) 2018) in order to minimize the number of migrants not being registered upon arrival. To better control migration into the European Union, they implemented pre-identification, registration, photo and fingerprint operations in the hotspots (Capitani 2016: 4, see the Introduction). Hence, since 2015, fingerprints have been registered as soon as the migrants disembark from the boats and are transferred to the camps created as part of the hotspot architecture. The tip of each finger is live scanned and stored in a database such as EURODAC as a digital image, as illustrated in Part I. By scanning the tip of individuals' fingers, the European authorities can thus gain access to the history of their past whereabouts. Not all the arriving migrants disembark in the geographically located hotspots, however. In such cases, a mobile hotspot team will be in charge of either registering the fingerprint on location or transporting the migrants to a location where registration can take place. The arriving migrants are not allowed to leave the hotspots before their fingerprints have been registered; this decreases the chances of avoiding biometric registration. In legal terms, this means that they cannot leave Italy to apply for asylum elsewhere, according to the Dublin regulation (see the Introduction).[5]

Many of the Somali women and men whom I encountered in Italy were keen to avoid the registration of their fingerprints there due to Italy's lack of economic stability. The absence of any future hope of economic and social prosperity turned Italy as a site of registration into a prison from which they could not escape and transformed their bodies into readable objects against their will. As a Somali man explained during an interview, 'The registration of our fingerprints is like experiencing judgment day while we are still alive' (see Chapter 6).

In Part III of the present book, the focus is on the role that the biometrically registered body plays among Somali women and men *en route,* showing how this manifests itself in different territorial sites as they move through the biometric border world.

Biometric registration and the ambiguity of care

Within the context of undocumented migration, biometric registration has been characterized as a form of control that hinders migrants' onward movements (Van der Ploeg 1999, 2003). In the literature, for example, by Watters (2007) and Broeders (2007), this form of control has been associated with a specific mode of distinction in the humanitarian care provided to migrants by the authorities. Watters and Broeders demonstrate how the state uses the biometric identification of asylum-seekers as a tool to provide care for those who are legally categorized as deserving it, such as those who are acknowledged by international conventions as refugees, while denying it to those who are not seen to be entitled to refugee status.

In Chapter 5, I show that as a form of control, biometric registration must be seen in the light of a complex mesh of contradictory care relations associated with transnational migration.[6] These relations include what I have called *supranational global care* practised by UNHCR workers towards refugees in a Kenyan refugee camp; *European humanitarian care* provided to refugees under the formal European asylum system; *transnational care* by Somali refugees of family members in the country of origin; and *interpersonal care* that can arise between two people, such as a Somali asylum-seeker and two Danish police officers, when one is in a desperate situation and the other(s) are not.

This does not mean that 'caring or being cared for is [. . .] necessarily rewarding and comforting' (Bellacasa 2012: 198–199). But instead of viewing the lack of comfort or reward through a moral perspective that focuses on the negative aspects or the absence of care, as is often done in good faith when portraying relations between a powerful state and its less powerful migrants, this chapter contributes new knowledge by arguing that the different forms of care point to the ambiguity of care that is involved when biometric registration is practised in the biometric border world. Care is never only negative, controlling or absent. Instead, it is the result of an intertwined duality of care (as comfort) and control.

In the field

> The increasing number of African migrants is clearly visible in some of the major Italian cities, which attract many hopeful young Africans who believe their chances of finding some sort of housing or work are better in the larger cities. Milan is one of the major Italian migration hubs, and this is where I encounter Fuuad in the spring of 2017, a Somali man in his early 30s and a friend of a friend. He has agreed to tell me about his experiences of living in Italy as a refugee. We meet in the square in front of the Central Station, a place that in many ways showcases Fuuad and other refugees' lives in Italy. A mixture of local Italians and tourists pass through the square on their way to the main tourist attractions or en route to other destinations. This form of mobility is what Fuuad and others dream about – being able to travel freely across geographical borders without being forced to return to Italy because this is where their fingerprints have been biometrically registered. It is in particular African, Asian and Middle Eastern men who occupy the square in front of the Central Station. Their belongings are packed in suitcases and large black plastic bags leant against the trees, while they themselves are resting in the shade or trying to sell small items to local Italians or tourists passing by.

For African, Asian and Middle Eastern men, the suitcases symbolize a life of immobility, even though they also enable them to become mobile when, for example, the police raid the parks or occupied buildings where the men sleep illegally.

As one of the founding members of the European Union, Italy's responses to migration flows and policies have been influenced by the EU for many years (Schuster 2005: 760). Italy used to be mostly an emigrant-producing country, but in the 1980s, this situation was reversed when the larger Northern European countries started to implement more restrictive policies towards immigrants (Schuster 2005: 759; Lucht 2012: 22). This meant that, all of a sudden, Italy had to deal with large numbers of non-Italians who were not only travelling through the country but were choosing or being forced to stay there, many ending up

Figure III.1 Suitcases in front of Milan Central Station, Italy, 2017
Source: Photo by the author.

as undocumented migrants, the so-called *clandestinos* (ibid.). In 2000, 85% of the immigrants in Italy were non-European, the majority undocumented. Many of them were attempting to move farther north, thereby converting Italy into 'a "backdoor" to Europe' (Lucht 2012: 22). Italy's way of dealing with this influx of people has been shaped by and developed alongside those of other EU countries, meaning that it has implemented many of the EU's migration management initiatives. From 1986 to 2002, Italy[7] sought to regularize illegal or clandestine immigrants in the country by means of amnesties, which were followed, however, by tightened laws against illegal immigrants involving additional visa restrictions and severe fines for being caught without document, and an increase in deportations (Schuster 2005: 760–761; Lucht 2012: 22–23).

The meeting with Fuuad in Milan was the culmination of almost a decade's work among transnational Somali refugees. Thus, I had 'snowballed' my way from Denmark to Ethiopia and Somaliland, then on to Turkey and Greece, and finally to Italy.[8] My connections among Somalis *en route* opened doors in every country I set foot in. I quickly learned to move in the social landscape according to Somalis' own practices, especially concerning the sharing of information. My frequent conversations with my Somali friends; their knowledge about the social, geographical and biometric landscape; and their introducing me to other Somali refugees added to my information about the perceptions, positions and practices of biometrically registered Somali refugees in Italy. I was denied access to the actual hotspots by the Italian authorities, but the sharing of information brought me to the churches, government housing centres, parks and occupied buildings where the Somalis would eat and sleep. I would interview them at their usual hangouts or talk through social media with those who had fled Italy in an attempt to seek asylum elsewhere. Following the information flows of Somali refugees in Italy also brought me in contact with those who were part of the Italian system of care for refugees. Police officers, social workers, forensic scientists, doctors and volunteers all played various roles in what motivated and produced the refugees' particular everyday decisions and practices.

Following these Somalis and the flows of information that were shared among them provided me with valuable insights into how they managed on a day-to-day basis. They had no master plan, nor did they constitute a coherent and close-knit network. They took small decisions every day based on whatever information they managed to glean from other migrants' experiences about the changing European legal framework and the lack of care within Italy. In addition, these decisions also depended on the Somalis' current social connections, economic possibilities and personal situations.

Chapter outline

Chapter 5, *'Fleeting biometric encounters: care and control at Italian border sites'*, takes its point of departure in an Italian hotspot where fleeting biometric encounters between newly arrived Somalis and Italian police officers take place every day. The chapter examines how the use of biometric technologies shapes different forms of care in the biometric border world and how control is an integral part of, and motive for, providing such care. It does so by zooming in on the registration of newly arrived Somalis' fingerprints and its consequences for Somalis as they leave the hotspots and move into the Italian system of refugee care.

Chapter 6, 'In-formation' *and* 'out-formation': *routines and gaps* en route' takes its point of departure in the stories of four Somalis who are officially recognized as refugees in Europe, three of whom have been biometrically registered in Italy through fingerprinting. The chapter shows how biometrically registered refugees in overburdened localities such as Italy attempt to compensate for the gaps in the care system through the search for *in-formation*, that is, for bits of knowledge which are not fixed but always *in formation*. This search for *in-formation* reveals the gaps within the biometric border world that might make it possible for Somali refugees to continue their search for a better life.

Notes

1 Eighty-five percent of the worlds' refugees are still hosted by developing countries (UNHCR 2017) and are thus not headed towards Europe.
2 Throughout Part III, the term *migrant* includes all people *en route*. The terms *asylum-seeker* and *refugee* are used when the legal category is relevant to the argument being made.
3 In 1991 Somaliland, formerly known as Northwest Somalia, declared its independence from the rest of Somalia. The Republic of Somaliland adopted its own constitution and political system and functions today as an independent state, although it is not internationally recognized (Hansen 2006: 9).
4 Al-Shabaab means 'the Youth' in Arabic. It arose in Somalia as a youth wing of the Union of Islamic Courts, which ruled Somalia in 2006.
5 'The right to apply for asylum is laid down in the Geneva Convention, which all 28 EU member states have signed and which has been incorporated into the EU treaties' (European Parliament). Norway, Island, Switzerland and Liechtenstein have also joined the Dublin regulation.
6 See also the work of Levitt et al. (2016), Dobbs and Levitt (2017), which explores contradictory care relations during transnational migration.
7 Although Italy is most often presented as one coherent unity, the country is very diverse internally in its approaches towards migration. There are different political wings and approaches depending on where and at which level one looks, for example, whether one is describing national, regional or local political approaches. The Italian population is also very mixed in its actions towards and perceptions of migrants located in Italy, some opposing migration, while others fight for migrants' rights.
8 The snowball method refers to the researcher conducting an interview with one or two key interlocutors, who will then introduce or refer the researcher to other interlocutors, who will do the same. The researcher will, in other words, be 'handed from person to person, and the sampling grows with each interview' (Russell Bernard 2013: 168).

5   Fleeting (biometric) encounters
Care and control at Italian border sites

In the summer of 2016, Italian police officers made Mukhtaar, a young Somali man who had just arrived in Sicily without valid travel documents, understand that if he did not register his fingerprints voluntarily, he would regret it. Faced with what he perceived as a thinly veiled threat, Mukhtaar complied and agreed to be fingerprinted, even though he realized that it meant he had to stay in Italy. The Italian police officer's practice of care towards Mukhtaar is an example of what I refer to as *European humanitarian care*. It is humanitarian because the Italian police officers represented not only the Italian but also the whole European asylum system, which seeks both to control and to provide humanitarian protection for those crossing its borders. The word *humanitarian* has a dual meaning. The *Oxford English Dictionary* defines *humanitarian* as concerned with reducing suffering and improving the conditions that people live in. It is, in other words, 'a concern for distant strangers' (O'Sullivan et al. 2016: 1). In the literature, however, being *humanitarian* is also described as a form of governance (Iriye 1999: 435) in a 'hierarchy of humanity' (Fassin 2010). It is the dual meaning of *humanitarian* care that is explored in this chapter.

Governance, or what I define as control, along with, for example, Van der Ploeg (1999, 2003), was practised when Italian police officers forced Mukhtaar to register his fingerprints through the threat or the use of violence. Biometric registration is a way to uphold European rules on migration management by permanently categorizing a person as eligible for either asylum or deportation by storing data about their bodies in a European database. Registering people biometrically serves the secondary purpose of making the process of providing humanitarian care faster and smoother. As an Italian human rights lawyer explained during an interview, registering new arrivals was the most effective way for the European asylum system to minimize the risks of migrants being trafficked or in other ways being taken advantage of in the underground economy. It also eased the process of providing legal housing, medical attention and locating lost migrants.

The newly arrived women, children and men did not necessarily experience such humanitarian care as rewarding or comforting.[1] In fact, Mukhtaar and many other Somali men and women perceived Italy's version of humanitarian care very negatively due to its limitations. As explained later in the chapter, the Somali refugees needed more permanent housing, more job opportunities and medical attention from the Italian state if they were to provide transnational care[2] for their family members in their country of origin.

Mukhtaar's fleeting biometric encounter with the Italian police officers at an Italian border point inspired the two questions I pose in this chapter. First, what forms of care and control are practised during encounters between European police officers, social workers and Somali asylum-seekers before, during and after the biometric registration of a fingerprint? Second, what role does the registration of fingerprints play in social relations involving care and control? In other words, the chapter examines how the use of biometric technologies shapes different forms of care in the biometric border world, showing how control is an integral part of, and motive for, providing such care.

The study follows three Somali men *en route*, focusing on the two biometric borders they encounter: their own bodies and the territorial locations they pass through or locate themselves in. It explores how biometric encounters between the representatives of various authorities, technologies and men and women *en route* are played out as they move through different stages of migration, from a refugee camp to an asylum process to acknowledgement as a refugee and finally deportation.

The contradictory care relations associated with transnational migration are defined as follows. Starting the chapter in a Kenyan refugee camp, I define care as *supranational care* because it is practised by UN High Commissioner for Refugees (UNCHR), an international and thus supranational organization. This care includes the biometric registration of refugees through fingerprints and iris scans to control fraud and ease the aid processes. This is carried out in cooperation with the Kenyan state. The second form of care explored in this chapter is the *humanitarian care* provided by the formal European asylum system to asylum-seekers and refugees such as Mukhtaar. The chapter does this, first, by further exploring Mukhtaar's encounter at this particular border site; second, by focusing on a meeting between an Italian social worker and a group of Somali refugees; and, third, by shedding light on the consequences of *humanitarian care* for Mukhtaar, namely life as a homeless person. Wedged into the *humanitarian care* provided by the formal European asylum system to asylum-seekers and refugees, the chapter also presents *transnational care* provided by Somali women and men *en route* for family members in Somalia. Care in this context means migration that can give them refugee status and provide them with a job, a roof over their heads and access to education. Achieving security for themselves in Italy will enable them to secure the livelihoods of their families at home, which will also improve their social position in their country of origin. Some may even be able to obtain family reunification and bring their family members to Europe (see Part IV). The practice of transnational care is what leads to the final section in this chapter, which describes *interpersonal care* during a flight from deportation. *Interpersonal care*, I argue, is what can arise between two or more people when one of them is in a desperate situation. This is exemplified by two Danish police officers showing sympathy towards a young Somali man and providing him with money when he fell sick and was in desperate need of a place to stay in Italy.

Supranational acts of care

Following the practices of Somali women and men *en route*, it becomes clear that the idea of 'a world of no borders and no political regulations' (as described in the introduction to this section) has all but disappeared. Territorial borders guarded by border patrols are rapidly being reinforced on the African continent, as well as in Europe. Somali women and men are today confronted with territorial borders as soon as they attempt to leave their country. Some Somalis described to me how they had migrated from Somaliland in the middle of the night, hoping they could cross the border undetected and travel northwards through, for example, Ethiopia, even though the borders were guarded by the border police. Others explained that they had planned that their initial steps of migration would take place at a time of major celebrations, such as Eid (in Somali *Ciid*), when border control was less stringent due to the high number of people travelling. Somalis are also confronted with body borders as soon as they enter one of the several refugee camps in the Horn of Africa, where they are registered biometrically through iris scans or fingerprint registration, and thus turned into readily readable objects of control.

By means of the UNHCR's use of biometric registration in a Kenyan refugee camp, this section shows how the fine line between assistance towards and control of refugees has become blurred with the implementation of biometric technology. It demonstrates that these practices of what I call *supranational care* involve an entangling of assistance, protection and migration management that mixes national and even continental interests. They thereby create a strong sense of 'risks stemming from successful uses of biometric technology' (Jacobsen 2015: 144) among the refugees on their very first steps *en route* towards Europe.

Created in 1951 after the Second World War, the UNHCR's main goal was to provide humanitarian assistance and protection to refugees in need worldwide. Since the 1990s, however, the organization has developed from focusing only on refugee protection to considering migration management as an activity complementing protection (Scheel and Ratfisch 2014: 925). The increased role afforded to control when providing assistance and protection within the UNHCR reflects a long-standing dilemma among politicians and practitioners over whether to prioritize practices of refugee protection (care as nurture) or national security concerns (care as control). In addition, as non-governmental organizations (NGOs) such as the UNHCR increase practices of control and management when providing assistance in refugee camps, the line between what count as practices of protection and practices of control becomes more fluid and blurred. This grey zone has become particularly visible since the beginning of the 2000s, when the UNHCR itself began to implement biometric technologies such as fingerprinting and iris scanning (Jacobsen 2017: 529).

According to the UNHCR, the implementation of iris scans and the registration of fingerprints in refugee camps are meant to optimize care and protection (Jacobsen 2015). Aid can be distributed more efficiently in refugee camps if biometric identification processes are used, and previous delays in the delivery of aid due to, for example, lost, forged or exchanged food ration cards or long lines of waiting can be avoided. The implementation of iris scans and fingerprint registration is, in other words, a way for the UNHCR to provide food items, shelter and other forms of aid to a great number of people in a short amount of time while at the same time serving as a tool against fraud (Jacobsen 2017: 537). This has been the case in Kenya, where many of the Somalis from South Somalia I encountered in Turkey and Greece had spent months, if not years, before fleeing towards Europe. In Kenya, the UNHCR's assistance towards refugees in the Dadaab refugee camp, which for many years was the world's largest, consists in providing aid while at the same time controlling those requesting assistance, as already mentioned. This is done by cross-checking the biometric data of every refugee registered in the UNHCR's database with the Kenyan state's database that stores the biometric data of Kenyan nationals (Jacobsen 2017). If there is a hit – a match – it means that the same person has requested care from both the Kenyan state and the UNHCR. The body thus becomes the most trusted source of information, but when 'letting technology loose in the wild' (see Part I) – a reference to the transfer of biometric technologies from the tech labs to border sites – the risk of technological failure arises. To be successful, the unique patterns on the finger are 'live scanned' into 'digital images through the sensing of the finger surface with an electronic scanner' of high quality (see Introduction to the book). These digital images are then most often developed into 'digital templates' that can be stored and/or compared, although not all digital images are of sufficient quality to be recognized as producing sufficiently good results and thus able to be used for storing and comparison. In migration contexts, this might be due to self-inflicted injuries to the fingertips or physical objection to the registration process,[3] but it can also be caused by sickness, previous manual labour or old age, which may make the fingertips of people on the move unreadable (cf. Magnet 2011; see also the Introduction to this book).

Fingerprint registration and iris scans, when successful, enable both the UNHCR and the Kenyan state to maintain a form of control over the high number of refugees, most of them from neighbouring Somalia, in the present and future because the biometric data of refugees is being stored without standardized protocols governing how and for how long it should be stored (Jacobsen 2017: 537– 538). For the Kenyan state, the information gained from the UNHCR's biometric database provides them with a tool to control the whereabouts of Somali refugees in the country, who, by virtue of their definition as Somali, are suspected of being terrorists. The Kenyan government, for example, has expressed concerns about refugees as potential terrorist threats (Jacobsen 2017: 541) due to several attacks on civilians in the country by the Islamist group al-Shabaab (Mutiga and Graham-Harrison May 2016; Weitzberg 2017: 5). As a result, the Kenyan government in 2016 threatened to close down one of the world's biggest refugee camps and to deport more than 330,000 refugees, most of them Somalis. Violent attacks conducted by Somali nationals have in many ways led any Somali to be perceived as a potential threat, not only in Kenya but worldwide (see e.g. Besteman 2016, 2017). First, this conflicts with the definition of Somalis as refugees in need of protection. Second, it shows how the UNHCR's practices of biometric registration and data storing might unintentionally turn the physical bodies of Somali women and men into potential figures of threat. As a consequence, Somalis located in Kenya experience more insecurity from not knowing whose interests are being served when they are biometrically registered in the refugee camps. Intimate data about every Somali in Kenya could be shared with Somalia by the Kenyan state, even though Somalia is a country the Somalis have fled from out of fears for their lives. Third, despite the fact that a Somali refugee is provided with assistance as a result of being biometrically registered, the aim being to avoid delays, some refugees describe the assistance they are given as insufficient and delayed, if it is provided at all. Subeer, a Somali in his late 30s, described during interviews and informal conversations how he had registered himself with the UNHCR in Kenya, which had acknowledged him as a refugee and approved him for resettlement. However, the assistance provided for him and his family was not sufficient: 'Hearing the little baby screaming, and you don't have anything'. This was how Subeer experienced the lack of assistance. In addition, after four years, the UNHCR had still not moved any closer to the resettlement process than when Subeer was first registered. 'Then I just give up and I come here', he said, *here* being Turkey, where I met him (Simonsen 2017). Hence, the experiences of supranational care in the Horn of Africa led some Somalis to continue their journeys towards what they hoped would be a more secure livelihood in Europe, despite an increased risk of facing territorial and biometric border control *en route*.

Following the 'bodies' of Somali women and men *en route*, Europe is where this chapter now turns. More specifically, the next section explores the way Somali women and men are provided with European humanitarian care by the formal European system for asylum-seekers as they turn from being undocumented migrants into asylum-seekers upon arrival in Italy.

European biometric politics of care and control

Relief. Happiness. Gratefulness. Hopeful. Hungry. Thirsty. Sick. Tired. These were some of the adjectives used by Somali women and men to describe how they had felt when they had just arrived at the Italian shore. Many of them described a tremendous joy at being rescued by the Italians, who would care for them upon arrival. Mukthaar had experienced an immediate sense of care when he arrived in Italy in the summer of 2016 after several failed attempts at crossing the sea. The boat he had travelled in was made of rubber, and when he and his fellow travellers came close to what they hoped were Italian soldiers, they made a hole in the boat to make sure that the Italians would help them – and they did. 'The Italians welcomed us, they protected us from the cold, and they provided us with food', Mukthaar explained. But for the majority of the Somalis among whom I conducted fieldwork, the sensational feeling of being alive and cared for in Europe was soon replaced by distress. Many were forced to have their fingerprints registered and, as a result, seek asylum in Italy. This was the case for Mukhtaar, who recalls the following episode in which he was turned into an asylum-seeker in Italy as follows:

I tried to refuse getting my fingerprints registered, and the policemen said, if you try to refuse again, you will see what we promised you . . . the words they were using, it made me scared. That is why I said, 'Don't lose your life, put the fingerprint', because six police officers tried to attack me when I was not ready to put the fingerprint.

At the reception camp where Mukhtaar had been taken upon his arrival, a Somali woman had approached him and advised him to register voluntarily: 'Please put your fingerprint, don't risk your life because these police officers also took my fingerprint by force, they beat me and other women'. The choice became one of obtaining immediate care for the sake of his own life and his family's future. Forced to have his fingerprints registered, Mukhtaar experienced first-hand what European humanitarian care could entail for asylum-seekers.

In the European asylum agency, undocumented migrants' concerns about the risks associated with successful uses of biometric technology and the consequences it can have for their futures are not given much attention. Instead, the focus is on making sure that the southern borders of Europe are maintained by means of fingerprint registration because it is here that the majority of migrants arrive. In 2014, the EU received a total of 626,715 asylum requests. Despite the fact that around 170,000 migrants arrived by boat in Italy that year, only about a third applied for asylum in the country. Countries such as Germany and Sweden received by far the largest number of asylum applications, with 202,815 in the former and 81,325 in the latter (European Commission 2015b). This can be correlated with the fact that the majority of the Somali women and men I encountered during fieldwork during that period made it out of Italy without having their fingerprints registered. Those who were caught by the Italian border police would narrate stories of pleading with the police to not register their fingerprints. They would tell the officers that they were searching for a better life and that they wanted to reunite themselves with family members living outside of Italy. As a result, many were allowed to continue their journeys without registering.

While the Somalis described the Italian police officers' goodwill in not registering them as a sign of solidarity with and care towards them, the latter's actions could also be interpreted as a way for them to take back control, according to Scherer (2015). They, along with the wider public in Italy (Frassoni, interviewed by Denkova 2016), felt that Italy had been abandoned by the European Union when the number of non-European migrants started to rise. By letting migrants cross its borders further into Europe, the police officers could re-establish some control in their own country, as this would minimize the number of migrants in Italian custody, as well as the economic burden of looking after them. Letting migrants cross was a way to implement a sense of control locally in a country in which the government was losing popular support because it could not convince Europe to practice care in the form of burden sharing (Frassoni, interviewed by Denkova 2016). Hence, despite framing the control and assistance of refugees in the European context as European humanitarian care, there is no single common practice or definition of what this consists of. Instead, it echoes Feldman's argument (2011b) that local actors practice the so-called migration apparatus in very different ways.

Member states' unwillingness to distribute refugees more equally within Europe was criticized by the European Commission itself. The president of the European Commission, Jean-Claude Juncker, urged the North European member state countries not to leave countries such as Italy, Greece and Hungary in the lurch with this challenge in his opening speech in September 2015:

> We now need immediate action. We cannot leave Italy, Greece and Hungary to fare alone. Just as we would not leave any other EU Member State alone. For if it is Syria and Libya people are fleeing from today, it could just as easily be Ukraine tomorrow.
>
> (European Commission 9th September 2015b)

At the same time, he explained that this would 'require a strong effort in European solidarity. Before the summer, we did not receive the backing from Member States I had hoped for. But I see that the mood is turning. And I believe it is high time for this' (European Commission 9th September 2015b). The change of mood resulted in a proposal to increase the number of relocations of people seeking international protection in Italy and Greece from 40,000 to 160,000 people (European Commission 9th September 2015b) but only for nationalities with an asylum acceptance rate of more than 75 percent within Europe. It also led to one major EU initiative in 2015, namely the creation of so-called hotspots. As mentioned in the introduction, the aim of the hotspot approach was to have a 100 percent identification rate and thus to be able to identify, screen and filter all newly arrived men, women and children through pre-identification, registration, photographs and fingerprinting (Capitani 2016: 4), and this is still the case. The hotspot, in other words, became a way to uphold the Dublin Regulation by turning the body into a border that could be located anywhere.

Hence, for the European Commission, as well as being a functioning relocation scheme, care was defined in terms of increasing the control of people arriving at Italian and Greek border sites through the use of biometric technologies. Care by and towards the European Union was also designed to make sure that such practices continued. The European Commission opened infringement proceedings against both Italy and Greece due to their failure to implement fingerprint registration in EURODAC (European Commission 2016), the database containing fingerprints of all asylum-seekers from 14 years of age and older (Schuster 2011: 404). The infringement proceedings and other migration management initiatives changed the practices of many local Italian police officers. In 2014, it was not a criminal offence for men, women and children who arrived in Italy without documents to resist having their fingerprints registered, meaning that Italian police officers could not legally force anyone to register (Scherer 2015). In 2015, however, the European Commission proposed 'a more solid legal framework to perform hotspot activities and in particular to allow the use of force for fingerprinting and to include provisions on longer term retention for those migrants that resist fingerprinting' (2016: 3). The new legal framework, which allowed a higher level of control over fingerprint registration by Italian border police, was made clear to Mukhtaar through a cultural mediator. The mediator translated the words of an Italian police officer when Mukhtaar resisted having his fingerprints registered. Mukhtaar recounts the mediator's translation as follows:

> If you don't put your fingerprint here, you will see some problem from us, because we have to keep our law, this is the rule and regulation of our country, if you don't put, also you will lose your life, this is our rules and regulation.

The words *lose your life* were a threat and did not mean literally that he would lose his life. It referred to a new situation in Italy in which in many instances the border police had gone from practising control by letting potential asylum-seekers such as Mukhtaar go without registering to a situation in which registration of the body had become the only way to manage migration within both Italy and the EU more generally.

Care for the family

Before moving on to the third stage – from asylum-seeker to refugee – it is important to shed light on why Somali women and men resisted having their fingerprints registered upon arrival in Italy. This section explores what I have called *transnational care*, that is, the care practised by Somali women and men *en route*, which often clashes with the European humanitarian care practised by Italian border police upon arrival.

I argue that Somali women and men resist having their fingerprints registered in Italy as an act of care towards their families. Migrating to greener pastures is the ultimate act of care when the geographical and socio-economic situations migrants find themselves in do not allow them to provide prosperity to the family. As explained in the beginning of Part III, people all over the world migrate to give themselves and their families more security. Due to the war and continuing instability in greater Somalia, Somalis could no longer migrate to greener pastures within Somalia itself, as there were none according to those with whom I conducted fieldwork. Instead, they had to leave Somalia, making their practices of care transnational. The majority of those migrating in the world, including Subeer and Mukhtaar, search for security in neighbouring countries. Subeer had fled to Kenya and Mukhtaar to Uganda but without any luck, which was why they decided to head towards Europe. But it had to be the right kind of Europe. Many had heard stories of how Somalis were homeless and received food from the churches in Italy, in contrast to Somalis in Northern Europe, who were provided with a form of European humanitarian assistance that enabled them to send money home or even to apply for family reunification. This was why Mukhtaar tried to avoid being registered by the Italian police officers. Recalling this episode, Mukhtaar explained:

> We were arguing, because the police officers said, 'You have to put your fingerprint'. I said I cannot because I did not ask you asylum here. They said 'OK, what do you want?' I said I want to leave from here, I want to go to

another country because I left my family in Somalia. They are sleeping outside in Somalia because they don't have anything. I want to help my family. That is why I don't want to stay here.

Biometric registration would, according to Mukhtaar, fix him in a particular time and place. As a consequence, he would be unable to create a better situation for his family or to live up to his obligations as the oldest sibling. At the same time, his decision to migrate was suffused with ambiguity. Mukhtaar had migrated as a way to provide security for his family because it was too dangerous for him to continue working in Somalia. His journey, however, was expensive; Mukhtaar's family had invested more than 14,000 USD to finance his journey to Libya, the last stop before Italy. As a result, his family was now homeless. Many of the families whose sons or daughters I conducted fieldwork with *en route* had sold everything they owned and, in addition, had borrowed money to pay for the journey to Europe. Hence, not only had they literally invested everything they had in Mukhtaar's journey, but this also meant that he was not there to take care of them.

Mukhtaar grew up with his father, mother and siblings. His father worked as a medical doctor with the African Union Mission in Somalia (AMISOM). AMISOM conducted the first peacekeeping operation in Somalia in 2007 (AMISOM) as an attempt to install peace in the country after the outbreak of the civil war in 1991 and the many years of continuing conflicts and unrest. Somalia, however, is still suffering unrest and conflicts between the government, its allies and insurgent groups such as al-Shabaab. On 28 February 2019, at least 24 people were killed and 55 wounded as a result of car bomb explosions followed by a day-long siege in the city centre of Somalia's capital Mogadishu. The deadly attacks by al-Shabaab was a reaction against the US military carrying out airstrikes in Somalia against al-Shabaab (France 24). Mukhtaar lost his father in the ongoing conflict, and the responsibility to provide food and economic stability to the family fell on him. He took over his father's job as a doctor but was threatened by al-Shabaab to stop working for AMISOM. Mukhtaar decided to change his profession for fear of ending up as his father had done. He started working for a dentist, while his mother worked as a carrier in the market until she was injured in a bomb explosion in the city centre. This put extra pressure on Mukhtaar to continue to bring in the financial resources to buy food and other basic necessities for his family. He eventually opened his own dental clinic but was again contacted by al-Shabaab, who had received information that his list of patients included soldiers. Al-Shabaab ordered him to stop treating government and AMISOM soldiers, but Mukhtaar refused. One day he was told to come and attend a patient in a rural area, but the trip was a ruse by al-Shabaab, who successfully kidnapped him. For 19 days, he was kept in captivity until government soldiers rescued him and other prisoners.

As previously mentioned, upon his release Mukhtaar decided to migrate to Uganda with the hope of finding a job and thus continue to provide for his family. Things were difficult in Uganda, however, and the move did not enable him to care for his family, so his journey continued first to South Sudan and then to Libya. Despite suffering imprisonment and heavy beatings in both South Sudan and Libya by various criminal gangs that controlled undocumented migration, Muhktaar continued fleeing. Migration became his only response to a very fragile situation and the only way to maintain some hope that he could assist his family to a more secure livelihood.

Care was therefore about moving. It became transnational because it, first, meant that Somalis had to cross national borders to escape the insecure situations that had prevented them from creating the kinds of future they imagined for themselves and their families. Second, the care they practised was transnational because it depended on the humanitarian care of another nation state. Muhktaar could not provide for the family he had left behind if he did not receive both international protection in the form of refugee status and economic and legal opportunities that he could transform into care for his family. In Italy, this was lacking according to the majority of Somalis I conducted fieldwork with. Thus, the chances of being able to provide transnational care for their families decreased profoundly when they were biometrically registered in the hotspots and, as a result, were moved from there to the camps as asylum-seekers, and thus into the refugee care system. As refugees they would encounter, among others, Italian social workers, who, as with the Italian police officers, practised European humanitarian care but in very different ways. For Alessandra, an Italian social worker, European humanitarian care in the form of biometric registration for purposes of identification was about creating a form of control that would enable her to provide care for refugees more smoothly. This was done by linking one body to one identity, as the following section shows.

Care in the Italian system for refugees: identifying the body

Italian social workers working with refugees were perplexed about the use of fingerprint registration at border sites in Italy. In their daily work, social workers provided housing, internships and language classes for those who had received refugee status in Italy and who had therefore been biometrically registered upon arrival. But they also encountered men and women who had avoided biometric registration in the hotspots and who had travelled to other locations in Italy. Alessandra, an Italian social worker whom I interviewed, stated in response to this situation, in which the large number of arrivals in Italy and the lack of registration meant that anyone could come and claim political asylum: 'It is positive that anyone can claim asylum because we are Italians. I think we welcome everyone, but it is negative because we cannot control everybody'. Italians have a long history of emigration and, in more recent years, immigration. It was to this experience, coupled with a generally open approach towards guests, that Alessandra was referring when saying that Italians welcomed everyone. However, she also thought it would be problematic if there was no way to register people because that would make her job of providing social services to refugees very difficult. Biometric registration was a way for her to provide services and help people because it provided her with the information she needed in order for her to determine what form of help, if any, should be provided. Alessandra exemplified what she meant by explaining how a Nigerian woman had come to ask her for help in locating her young daughter, with whom who she had lost contact after arriving in Sicily. Neither the woman nor her daughter had been registered by fingerprints upon arrival, which made it very difficult for Alessandra to locate the little girl, who was not in possession of any form of identification. Alessandra also argued that the storing of biometric data and access to it would minimize the time needed to provide information to the many rejected asylum-seekers who came to her office asking her to explain why their claims had been rejected. Currently, she had to arrange and schedule visits with the police to get access to peoples' biometric identification data, which could sometimes be a protracted process.

Identification, however, was not always straightforward. First, many asylum-seekers, like Mukhtaar, refused to be biometrically identified, which prolonged the process and exposed them to violent threats or even actual assaults. Second, the majority of Somalis who fled had done so without valid documents, as it was more or less impossible to travel to Europe on Somali documents. This meant that they would either travel without documents or that they would provide names and European documents that were either forged or had been borrowed *en route*, as earlier mentioned. The changing of names and documents at times made identification complicated for social workers like Alessandra. This was especially the case if the refugees had provided a different name than their own for their initial biometric registration. Many of the Somalis I encountered hoped to continue their journey out of Italy, and some therefore chose to give other names, surnames and dates of birth than their own when they were registered by Italian migration authorities. Alessandra described an episode involving a group of Somalis living in a camp for refugees who had had difficulties remembering which name they had provided when they registered:

> Two days ago, I was in a camp for refugees and I asked a group of Somalis 'Please can you fill in the form?' Nobody was able to fill in the form. I said, 'But can you write?' 'Yes, we can'. 'So write your names, date and place of birth'. But they couldn't. 'Why can't you do this?', I asked them.

Alessandra continued the story by answering her own question: 'because they forgot the name they said in Italy for the identification'. Alessandra continued: 'So I said, "OK go to your room, come here with the paper that Italy gave you and copy exactly your name and surname"'.

Out of eight people, only one wrote down his name right away. The others went back to their rooms to see the names they had written on their temporary permits to stay and then copied everything from these documents on to the forms that Alessandra had asked them to fill out. By letting them write down the names that corresponded to those they had already given, Alessandra could make sure that these men could become a part of the national system of refugee care because, as she argued:

> When they are refugees, it is not a problem. Even if they change completely the name, it's fine. It's fine because it doesn't matter for the Italians, for

Europe; your name is X, OK, then stay forever with that name. The problem is during the process, if they want to change the name, then it's a problem. The last possibility they have to change the name is in front of the Commission. When in front of the Commission, they can say, 'When I was in Sicily I said that my name was Hamed. I'm so sorry. My real name is Mohammed'. It is the only solution they have to change their name. After that, you have to make an appeal to the high court. It starts to become a big problem because they want the birth certificate and so on.

Alessandra's efforts to provide care as a social worker focused on adapting the practices – in this case, those of Somali refugees – to those of the Italian system of care, which was based on linking a *unique body* with *a name*. Hence, the ambition was not to control whether the name provided was the birth name but, rather, that *a name* was provided that simply corresponded to that given at registration which would thus ensure a stable though possibly incorrect identity. In sum, Alessandra's practices of care were a result of navigating between the legal and administrative systems of welfare in Italy, her own personal interpretation of how care should be practised and the needs of the refugees. Navigating these different forms of care would enable Alessandra and her colleagues to provide social welfare. In contrast, many of the refugees described the European humanitarian care provided in the centres for refugees as inadequate. This feeling only increased when they had to leave the refugee centres after six months and more or less make it on their own in Italy, with nothing more in hand than a document stating that they had been recognized as refugees under the 1951 Refugee Convention. The next section turns to the experiences of receiving European humanitarian care as a refugee in the refugee centres and of living as a homeless person on the streets of Italy, towards what Mukhtaar described as 'the jungle'.

Life as a refugee in Italy

Being recognized as a refugee meant that Mukhtaar and others could make use of the national system of care. This system was implemented by social workers like Alessandra, who, with help from EU funding, provided refugees with a place to stay in the so-called SPRAR (Sistema di Protezione per Richiedenti Asilo e Refugati) centres. Here, people who had been granted asylum would be able to stay for around six months. Some could stay a bit longer if they had special physical or mental issues or conditions. During these six months, they would receive tuition in the Italian language, a place to sleep and food, and some would be offered internships of various sorts with Italian companies. The problem, however, arose in transferring them between the centres where they stayed during their application process, 'the CAS (Centro di Accoglienza per Richiedenti-Asilo) centres', and the SPRAR centres. Refugees ended up having to wait for a free bed in a SPRAR centre while being asked to leave the CAS centre because they had now acquired the status of recognized refugees. Some of them therefore ended up homeless. In addition, when the six months in the SPRAR centres were up, many returned to the streets.

Stories of Somalis sleeping in parks, spending their days strolling along the streets and only eating food that was provided free of charge by the churches were well known in Somalia. These stories were widespread because many had friends, family members or acquaintances who had had such experiences. Mukthaar and many of the other Somalis with whom I conducted fieldwork were among those who had been sleeping in parks and other areas at night. In the mornings, they would sit in different parks, stroll around and, when they felt hungry, head to a church, where they would register their names. Here, they would also register for a shower once a week. Mukhtaar described the nights in the parks as sleeping in 'a jungle'. Sleeping in a jungle meant that no laws existed and that people would fight, making it a very insecure place. And, as Mukthaar explained, 'no one can help you here in Italy. If you need food, someone will guide you in the direction of the church. If you need a place to sleep, another will guide you in the direction of a park – the park is our hotel', he said. At the time of our interview, Mukhtaar had managed to put a roof over his head, but only for two or three weeks. His family in Somalia knew a man who had a friend in Italy who had agreed to let him stay for a few weeks. After this he would again be left to the parks until the SPRAR centre could accommodate him as having obtained refugee status.

Living as a biometrically registered refugee in Italy was not only a form of legal categorization in a system that could not cope with the sheer numbers of refugees. It was also very much a social category that for many of the women and men I interviewed was very different from what they had experienced in Somalia. As Mukhtaar said,

> When I was in Somalia people respected me because I was someone official and I have confidence being a Somali living in my country. I don't live outside, I don't look for food, but when I come here, I experience all of this.

Some compared their status in Italy to that of a dog, arguing that the Italians cared more for their dogs than they did for them, with similarities to the situation of legal outlaws in Italian society (see Lucht 2012).

Life as a biometrically registered refugee in Italy, as well as the fear created by being in this situation, led some to try to escape Italy, although they were fully aware that there was one border that they could not fully erase: that constituted by their bodies. It is towards a failed attempt to escape Italy that the chapter now turns.

Deportation: Ali and the Danish policemen

Deportation is part of what characterizes the European authorities' practices of humanitarian care. It is an act of control, a demarcation of power and, for the Somali refugees, a symbol of failure. At the same time, it is in situations such as these that the double-edged characteristics of care can take shape in unexpected ways. The following section describes the encounter between Ali, a Somali asylum-seeker, and two Danish policemen as one where *interpersonal care* arises.

Interpersonal care is the practice of one human being providing immediate assistance to another in the form of either material objects or affection beyond what is required of the former as a professional.

Ali is a Somali man in his late 20s. He gets by, day by day, by sleeping in the parks and eating in the churches. He is, just like Mukhtaar, a recognized refugee in Italy. We sit in a corner of an outdoor café. Migration, Ali says, is like an education. 'You learn from crossing many countries. To migrate is like taking a Master's at a university'. He shares his main insights from his education as a migrant with me:

> Europe is two parts. Rich European countries and poor European countries. Mostly, poor countries are very racist. You are coming to Greece, you are coming to Serbia, you are coming to Hungary, and you don't have anything. But if you are coming to the rich countries, like Germany, Scandinavia, Denmark, Sweden, it is very beautiful [helpful] people, who want to help you.

As a master's student would do, Ali backs up his claim with data. The data contain his migration history from Italy to Denmark and back to Italy again. After having his fingerprints registered upon arrival in Italy, but before seeking asylum, Ali decides to flee Italy towards Northern Europe. However, he is detained at the Danish border by the border police, who run his fingerprints through EURO-DAC and find a match. Ali's fingerprints are recognized in the database as already being registered in Italy. Ali is surprised and angry because he has fled Italy with the hope of living a more secure life than that he experienced in Italy. The Danish police provide him with food, clothes and a ticket to go to a refugee camp for newly arrived asylum-seekers. From there, Ali is transferred to another camp. Ali describes the nights at the camp as good but also worrying, because he is in constant fear of being returned to Italy. He decides to leave the centre and goes underground. For Ali, this meant living with friends and family elsewhere in Denmark for the next one and a half years. It also meant living on the margins of society, not being able to take a legal job or educate himself. His main goal was to avoid any contact with the authorities. However, he falls ill from worrying, not eating and not sleeping well. Ali is hospitalized for 20 days, after which the police arrest him. Along with his medicine and his tablets, he is put in jail before being forcibly returned to Italy, escorted by two Danish policemen.

Listening to Ali and his migration history, I am struggling to understand how he has come to the conclusion that those who assisted him in Denmark are good people. What I hear, on the contrary, is a story of living on the margins of Danish society because of a biometric registration. The consequence of this is that he became physically ill, which in the end landed him in jail, despite his illness, and returned him to a life on the streets in Italy. But Ali insists, continuing his story: 'Three months I'm in prison. After three months, the police bring me all my tablets because I have to take my tablets for almost six months'. Two Danish policemen bring Ali the remaining tablets while escorting him on a flight back to Italy. He explains how he is very sick at the time but how the two policemen help him at the airport in Italy and how they have conversations with him. 'They are the best guys, the police of Denmark', he concludes and explains what they have told him during their conversation while sitting in the airplane:

> 'We are sorry. We like you to take a good life, a life in our country, but the system of Europe says that if you take the fingerprints of another country you must be returned. That is why you come back to Italy, but we would like you to have a good life in our country. We are sorry. The European Union has rules and regulations. All the rules of Europe are under the law, you know. That is why you must come back to Italy, but we are sorry'. When we came to the airport in Italy, I remember the police of Denmark give me 200 Euros and said to me: 'This is our own money, and now you can eat something for a few days'.

Despite the fact that the policemen's main task was to deport Ali to Italy against his will – a rather negative consequence for Ali – the situation developed into a form of inter-personal care to which Ali could relate. Besides upholding the European asylum system by enforcing imprisonment and deportation, the two policemen separated the migrant as a legal category from the migrant as a person, thus allowing a degree of intimacy to develop. The rather short trip in the airplane changed Ali from a biometrically registered body belonging, according to European law, in Italy into Ali, a person whom the two Danish policemen wanted to succeed in life. They showed this through their affection and financial gift. That was what made Danish people 'very beautiful people who want to help', according to Ali.

It was not only the personal contributions of the Danish policemen but also the professionalism of the Danish health care system that Ali defined in terms of good people wanting to help. The three months he was hospitalized and the provision of tablets to him that followed were practices of great care according to Ali. Recalling his time there, he explained:

> Every time I remember this hospital because it's the best hospital and I want to say 'tusind tak' ['thank you'] because they give me all my tablets. When I come back in Italy, I go to the hospital, we check my health, but they don't give me my tablets. That is why to say: thank you very much to this Danish hospital.

Coming from Denmark myself, I promised Ali that, should I ever go there, I would take his message to the staff, but to my great surprise I found that he had visited the hospital himself: 'I go with flowers and the paper of 'thank you so much'. Asking Ali how the staff had reacted to his kind gesture, he replied instead: 'You are very, very, good people – because if the hospital did not give me full tablets, still I'm sick because the hospital in Italy don't give me the tablets'.

In an attempt to conclude this story of Ali's appreciation of this particular hospital's service towards him – a service that all other ill patients receive as well – we might say that his conclusion has to be understood in light of the position he found himself in the EU. Like Hanne Mogensen's findings among HIV-infected Ugandan women living in Denmark who found themselves very lonely and excluded – un-integrated in many different ways (2011: 208) – but who still all spoke very warmly of everyone in the Danish health care system, I argue that Ali's lack of adequate health care in Italy led to his greater appreciation of the hospital staff in Denmark. This form of care put him in a position that was, in Mogensen's words, 'close and yet profoundly distant' (ibid.: 219). When Ali was arrested at the hospital, this meant that they could no longer care for him and could not visit him or call him, but according to Ali, they had nonetheless extended their practices of care by providing him with the tablets that would help to heal his body – a body that had betrayed him. First, his fingerprints had revealed his previous stay in Italy. Second, he had fallen ill – so ill that he had to be hospitalized. As a consequence, he was returned to Italy. But despite this, and despite the overarching system of control that had been incorporated into his physical being, he had experienced *interpersonal care*. Care can arise in the most unlikely places if time allows – in this case during a deportation.

Conclusion

This chapter has identified four different forms of care as migrants *en route* move from refugee camps in the Horn of Africa to become asylum-seekers in Italy, becoming recognized refugees who then attempt to continue their migration, which, in Ali's case, ended in his deportation back to Italy. What has been argued throughout this chapter is that the four different forms of care (supranational care, European humanitarian care, transnational care and inter-personal care) that are practised by different actors in very diverse ways and contexts arise in part as a consequence of the biometric registration of migrants' bodies – in this particular case their fingerprints. The chapter has also shown how practices of care in its different forms become intertwined with practices of control and vice versa by following the site of the moving body. This provides us with a more nuanced picture of the biometric border world – of the interaction between the authorities, biometric technologies and migrants in migration contexts – by illustrating that the care we as human beings perform is very diverse depending on the time, the context and the position we find ourselves in. It is never just about control or only about providing aid or other forms of assistance. In the biometric border world, it is always both.

These two arguments originate in my initial exploration of supranational care: the UNHCR's implementation of iris scans and the registration of fingerprints in the Horn of Africa. Here, it became clear that the success of biometric registration in this particular context poses as much of a risk to the Somali refugees as it represents a simplified avenue to receiving aid. The increased risks led some to flee towards Europe, initially arriving in Italy, where they went from undocumented migrants *en route* to asylum-seekers. Here they were fingerprinted and thus confronted with European humanitarian care at the Italian hotspots, at the centres for asylum-seekers and refugees and in their lives as homeless refugees on the streets of Italy. In the hotspots a *body* was matched with *a name*, which then restricted the movement of this particular body, decreasing the opportunities for Somali women and men to practice *transnational care*, that is, to provide for the families they had left behind, whether through financial aid or by applying for family reunification. Despite the contrast in caring practices – in viewing biometric registration as a tool to provide care for refugees versus a tool that restricts refugees from providing care for their families – the chapter ends by showing how two seemingly contrasting practices of care and control between two Danish policemen and a Somali asylum-seeker developed into an inter-personal form of care. In this situation, Ali's less favourable position, enforced by the two policemen, was put aside, at least until the flight ended. Instead, the distance in power and position developed into a form of physical closeness during the flight that drew them a bit closer to Ali as human beings.

Notes

1 See the work of Bellacasa, who argues that care is not necessarily experienced as rewarding or comforting (2012: 198–199).
2 *Transnational care* is defined later.
3 Some of the Somalis among whom I conducted fieldwork in Italy would deliberately burn the surface of their fingers, making it impossible for the authorities to scan their fingerprints, at least until the finger had healed.

6 *'In-formation'* and *'Out-formation'*

Routines and gaps *en route*

One warm sunny afternoon, Fuuad and I meet at the cafeteria that has served as our meeting spot in Milan throughout the summer. It is busy as usual, being visited by local Italians, tourists, refugees and an anthropologist. I have come not only to say my goodbyes to Fuuad, as I will be returning to Denmark, but also to arrange for my next trip to Italy. Fuuad, himself a recognized Somali refugee in Italy, has worked as my research assistant and thus assisted me in gathering information about the lives of Somali women and men who have been biometrically registered by being fingerprinted and given refugee status in Italy.

For Fuuad, life as a refugee in Italy has proved difficult. Reflecting upon the time he was given refugee status in Italy, he echoes the same experiences as Mukhtaar and Ali (see Chapter 5) of homelessness and obtaining food from the local Italian churches.

According to Fuuad, 'Italy only takes the fingerprint', meaning that Italy registers and provides documents but nothing more. And, as Fuuad notes, a piece of paper cannot bring food to the table. Fuuad continues by explaining that the EU has provided Italy with financial support to cope with the number of refugees in the country, but that it is not enough: 'They don't give house, job, nothing, not even education or training of some sort. You come from the boats and then directly to the streets'.[1]

The streets, defined by Mukhtaar as a jungle with no form of security (see Chapter 5), is not what Fuuad or any of the other Somali refugees have dreamt about. In their attempts to avoid life as a homeless person in Italy, many of them seek asylum elsewhere in Europe. Seeking asylum in a country other than the one they first arrived in violates the Dublin regulation (see Introduction) (Brekke and Brochmann 2014: 147; Fratzke 2015: 1). Nevertheless, Fuuad has previously attempted to do exactly that, but his fingerprints were recognized in EURODAC, and as a result, he was returned to Italy. This makes Italy feel like a prison in which he has been incarcerated, but without knowing by whom, as he puts it.

As we are wrapping up our conversation, trying to arrange what work lies ahead of us when I return to Italy, Fuuad tells me that he is planning to seek asylum in another European country. He will travel there to have his fingerprints registered, seek asylum, and then go back to Italy to await the further

Figures 6.1 and *6.2* Parks in Milan, Italy. Author took both pictures, summer 2017 in Milan, as she visited these parks in the company of Somalis who were, or had been, sleeping there at night

Figure 6.2

proceedings. Fuuad has a friend there, so he will give his friend's address to the authorities, and if they send him a letter or ask him to show up, his friend will alert him, and Fuuad will return to this country. I am rather surprised by this news from Fuuad, as he has just told me of his previous failed attempts to leave Italy. In addition, we have done numerous interviews together that tell the same story of Somali asylum-seekers and refugees from Italy whose fingerprints are recognized in EURODAC and, as a result, are returned to Italy by the authorities. My surprise shows on my face, and Fuuad continues. He explains that he will go to this particular country because he might be lucky, meaning that maybe they will give him refugee status: 'This country is a great country, so maybe, even if they see that I am registered in Italy, they don't care because they know that the conditions in Italy are difficult. I know friends who have experienced this'. Fuuad, in other words, is searching for gaps in the European asylum system, not as items or practices to be closed but, rather, as 'glimpsed openings in the closures around them' (Richter 2018: 27). I ask how Fuuad will manage to cross Italy's borders, referring to a story I heard the day before from another Somali who did not succeed in making it out of Italy. Fuuad answers that he knows a person who has just crossed that particular border one or two weeks ago, so it depends which documents you have and whether you have tickets to travel further. 'It depends if you have good documents or not', he explains.

It is clear from Fuuad's and other, similar accounts collected from Somali women and men in the course of migration that decisions concerning if, when, where and how to move forward very much depend on bits and pieces of information gleaned from the stories of other Somalis. It also appears that one successful story tends to cancel out the many negative ones. Building on this, the chapter discusses the search for *in-formation* (bits of stories and/or rumours that are always *in formation*; see the next paragraph) about routines and gaps in the biometric border world that can open ways to move on. It does so by exploring the ways in which two Somali women, Riyaan and Ladan, and two Somali men, Fuuad and Ali, base the small decisions they make about if, when, where and how to move on from the *in-formation* they have obtained. The chapter argues that biometrically registered refugees in overburdened localities such as Italy attempt to compensate for the gaps in the care system by searching for information. At the same time, however, this search for information reveals the gaps within the biometric border world that might make it possible for Somali refugees to keep moving in their search for a better life. In other words, this chapter shows how biometrically registered Somali refugees navigate the flows of *in-* and *out-formation* (information that the body releases that has been collected and stored; see next paragraph), which are constantly circulated to, by and about them in the biometric border world.

In-formation and *out-formation*

The Somali word *xog* is associated with the detailed information you need to make a decision. *Xog* thus implies bits and pieces of knowledge, clues and hints that you search for every time you find yourself in a situation in which you cannot make a decision without investigating the case further (Simonsen 2017). The biometrically registered Somali refugees in Europe I met were constantly searching for bits and pieces of *in-formation* that could guide them in their further journeys. At which borders were the risks of being compelled to have your fingerprints registered the highest, and what were the consequences? Which documents did they need to pass which territorial borders? Which new laws were being introduced in the EU, and what did that mean for their current and future moves and whereabouts? When was the best time to travel during the day and the year? How should you dress? Were there ways to avoid being registered by fingerprints? What should you say if you were caught? Finding answers to such questions was often considered a matter of life and death – as a Somali proverb says, 'Not having information is like experiencing hunger'.

These bits and pieces of information were very diverse in their content, form and shape, and the information collected was always *in formation*. I therefore use the term *in-formation* to refer to the process of searching for *xog*. It entails the circulation of hints, of bits and pieces of stories, rumours and news among close family members, distant relatives, friends located in Europe or Somalia, the friends, acquaintances and friends of friends whom the Somalis had come to know during *tahriib*, as well as the *muxalas* (the Somalis who had made it their profession to facilitate further journeys, most often for a fair amount of money). *In-formation* was what connected these very diverse groups of people. *In-formation* also refers to the content itself that had not yet turned into detailed knowledge (*xog*). Sometimes it was collected from the main source, meaning one who had experienced crossing a particular border or having his or her fingerprints registered in a hotspot in Italy. Or it might come from the latest news on the radio or TV. But just as often the *in-formation* gathered was not verified by a main source. Instead, it consisted of the stories of others, retold by being passed through many mouths and interpretations before ending as a form of *in-formation* for people such as Fuuad. In the academic literature, such *in-formation* has been defined as rumours. Among the Somali women and men *en route*, rumours were part of the *in-formation* they gathered and were used to exemplify and deal with the uncertainties that are constantly experienced *en route*. Despite the fact that this form of *in-formation* did not have a main source to verify it, for the Somalis it still played a positive role. It kept the future uncertain, meaning that it opened up the possibility of actually succeeding (Møhl 1997; Harney 2006) in, for example, applying for asylum elsewhere despite being biometrically registered in Italy, because this non-verifiable *in-formation* erased the certainty of failure. It provided a preliminary for one's next move, told by one Somali to another, about conditions outside the smaller group (Harney 2006: 377). In this way, 'rumour acts as improvised news in the absence of more formal and verifiable news. It offers an interpretative frame for those participating in its circulation' (Harney 2006: 376). *In-formation*, in other words, refers to both the process of collecting and circulating it and its content, which is not static and is often not verified by the main source.

Although they spent their days investing the majority of their time in collecting the latest *in-formation* in order to improve their livelihoods, the Somali women and men were not in control of the flows of what I call *out-formation* that were already in circulation. *Out-formation* I define as the signals, flows, elements, beats, rhythms and so on that their body releases and which are collected and registered in the biometric border world. It is the fingerprint registration popping up in EURODAC when Fuuad attempts to circumvent the Dublin regulation, it is the inability to receive medical care as an illegal asylum-seeker at a state hospital without being arrested (see Chapter 5), and it is one's identification in terms of certain categories, statuses and ID documents, such as an asylum-seeker's card versus a European passport.

The following empirical examples show how Somali asylum-seekers and refugees seek to navigate between the flows of *in-* and *out-formation* in the European biometric border world, as well as the small decisions that result from this in their everyday lives. I am interested in how these bits and pieces of *in-* and *out-formation* are evaluated, shared and used in practice.

When information travels fast

The majority of the Somalis I encountered during fieldwork had been *en route* for many years, and entering Europe did not necessarily change that. Instead, it made migration even harder, as it was not only territorial borders but also their own bodies as a border they were up against. One of the ways in which they sought to overcome these barriers was by constantly searching for *in-formation* that could guide them in which steps to take next. Following in the footsteps of Riyaan, this section explores what *in-formation is* among Somali women and men *en route*, from whom and where it is gathered, the constant process of *formation* it involves and the *out-formation* migrants seek to navigate within and around.

My first encounter with Riyaan was in Turkey in 2013. She was a warm, loving woman; a mother; a wife; and now also a migrant. She had fled Somalia due to threats made against her by al-Shabaab. She managed to flee Turkey, and we met again in Greece in the summer of 2014. It was not until we met in Greece that Riyaan started to worry about 'having her fingers taken', a reference to having her fingerprints registered biometrically in locations such as Greece and Italy. Inhabiting social, political and economic landscapes like those in Greece and Italy, which were well known for their lack of economic prosperity, led to a fear of forcibly being turned immobile. In 2011, however, the European Court of Human Rights overruled Greek asylum policy in a specific case due to the inhumane conditions for asylum-seekers in the country and the lack of a functioning system. This meant that the Dublin Regulation no longer applied to Greece (Topak 2014: 820). This made it possible for Riyaan, along with every other potential asylum-seeker, to attempt to seek asylum elsewhere in Europe if they succeeded in leaving Greece, which, however, was not easy without valid travel documents. Riyaan fled by boat out of Greece and into Italy, where the Italian border police caught her. Riyaan begged the Italian police officers to let her and the rest of the group she was with go without registering their fingerprints. They had families elsewhere in Europe, and they wanted a good life, she pleaded. She succeeded and managed to flee to a Northern European country where she was biometrically registered and applied for asylum. But her joy at this was short.

Riyaan called me on the phone one afternoon in the fall of 2016, crying. The authorities had rejected her case for asylum, and she was devastated. We decided to start searching for *in-formation* that could somehow guide her in how to respond to the rejection. I approached lawyers asking for information about the asylum laws. Who could represent her? What, if anything, could be done for her in the way of an appeal? I also contacted non-governmental organizations (NGOs) that specialized in working with rejected asylum-seekers and asked for their advice. Riyaan, on the other hand, sought information from other Somalis who were located in other Northern European countries. Speaking on the phone regularly, she told me the story of two Somali men who had done what she was hoping to do, namely to leave the country that had rejected her and instead seek and obtain asylum in a nearby country. I was worried for Riyaan should she decide to try to defeat the system and told her to stay put until I had gathered more information about her legal options. But Riyaan decided differently, and the next time I called her, her phone number no longer existed. She had left.

In my search for Riyaan, I did what I had seen so many of my Somali friends do when they were trying to locate someone. I called the Somalis I knew who also knew her from our time in Turkey and Greece, and after a while, I managed to locate her. She had made it and had applied for asylum again. For a while, both of us thought that she had succeeded in finding her way out of an otherwise hopeless situation. However, although she had managed to leave and seek asylum elsewhere on the basis of her constant conversations with other Somalis, she was not in control of the flows of *out-formation* that were already out there. Her fingerprints matched the EURODAC database, and consequently, Riyaan was returned to the northern European country she had recently fled from, leaving her devastated again. The story of the two Somali men that had served as an inspiration had backfired. She was not able to circumvent the physical border she was carrying around on the tips of her fingers. But Riyaan did not give up and did not accept the decision set out for her by the European asylum system. Instead, she continued to seek *in-formation* from other Somalis such as family, friends, acquaintances or complete strangers. They had either navigated in this system or knew of people who had, and this guided Riyaan in making her next decision, namely to return to Italy.

For Riyaan, it was information about Italy's struggles to live up to the time limit specified within the Dublin Regulation that enabled her to create a gap and challenge what was otherwise meant to be a routinized system of border work, despite her biometric registration in Northern Europe. The time limit works as follows. If, for example, Germany requests Italy to take charge of an asylum claim, the request has to be made by Germany within three months of the application, and Italy has two months to respond. If Italy does not respond within this period, it will be deemed to have accepted the asylum-seeker (Citizens Information Board). While Riyaan had obtained this *in-formation* due to the experiences of other Somalis *en route*, I, too, had gathered it from Italian lawyers, social workers and

NGO staff during interviews and informal conversations. They all explained to me that Italy's old bureaucratic and financially stretched asylum system did not have the resources to process the requests within the time limits set out.

Hence, instead of begging the Italian border guards to let them pass without being registered, some asylum-seekers such as Riyaan now started to return to Italy and intentionally seek asylum there. The reason was that they could at least obtain documents in countries such as Italy and Spain. Riyaan had found a gap in the system that had allowed her to defeat both the territorial and bodily borders that applied to people like her, and she was granted asylum in Italy. She was able to implement and adjust the *in-formation* she gathered much faster than the European system of asylum could.

Whereas Riyaan managed to locate a gap in the biometric border world through the *in-formation* collected from fellow Somalis, the following case shows Ladan, a young Somali woman, deciding to discard the *in-formation* collected among fellow Somalis. Instead, she gathers *in-formation* from an employee at an embassy, which enables her to negotiate the *out-formation* that is already registered about her in one of the most routinized and controlled areas of border control – the airport.

The airport: from *out-formation* to *in-formation*

It was midday when I met Ladan at a railway station in Denmark. We had not seen each other since we had met in Greece in the summer of 2014, where she had lived as an undocumented migrant. Like Riyaan, Ladan had left Greece because of its lack of economic stability and was able to seek asylum elsewhere due to the lifting of the Dublin Regulation as it applied to Greece. Now she was applying for asylum in another Northern European country and had decided to spend a holiday with her husband, who was located elsewhere in Europe. The holiday had just ended, and Ladan had jumped on to a train that was passing through Denmark, which was where we met.

I took Ladan out for dinner. We talked about everything and everyone before heading back to the bus, for which she had bought a ticket to take her back to the country in which she was currently seeking asylum. After getting her luggage into the bus and standing in line, Ladan was ready to embark on the trip. She showed her ID, a card showing that she was an asylum-seeker. But Ladan was refused entry on to the bus. The bus driver explained that she was not allowed to travel on an asylum-seeker's card, and if he were to let her on the bus regardless, he would risk a fine of around 50,000 DK. Besides not fully understanding the legal arrangements for asylum-seekers in Europe, Ladan had also not taken into account the fact that, during her holiday, obligatory ID control had been implemented in most of Northern Europe. The border control was first introduced by Sweden. This was followed up with ID control on 4 January 2016, originally intended to last for about four months but which in May 2016 was extended for another six months. This border and ID control affected everyone working in transportation, namely train, bus, ferry and airline companies, who because of the principle of 'carrier responsibility' were forced to implement the new, increased levels of security to avoid fines.

Ladan cried. I tried to calm her but panicked somewhat inside. Where should we go? The bus driver advised us to contact the embassy of the country she was seeking asylum in or to go to the police station, but as it was 10 p.m., we had no choice but to go back to my house. Ladan cried as I explained the situation because she, like me, was scared that she could not return and that her case would be adversely affected not only by this unauthorized movement but also because the illusion of reaching Europe as the answer to all her problems had been torn violently apart. Finally, Ladan cried because she realized that she could not see her husband again. This was a form of prison in that she again found herself not free to move, as Fuuad described Italy. When we came home, Ladan started to call every Somali she considered to be a source of *in-formation*. This continued for the 14 nights she ended up staying in Denmark. She talked throughout the night, hearing the stories of other Somalis who had been in similar situations to her own or who had heard of others who had. Many times during the day, Ladan would suggest a new way of moving out of Denmark to me on the basis of new *in-formation* she had obtained. I would seek to verify the information from legal sources, and we would discuss the various options that she was presented with. Although the bits and pieces of *in-formation* would give her new hope, they also made her more confused. Some suggested trying her luck on the train or bus or being smuggled by car; others suggested buying fake IDs and taking her chances with those. Meeting a dead end each time, Ladan finally decided that she would go to the embassy, and thus, we went there together. The embassy asked her to fill out a piece of paper in which, among other things, she had to state why she had left. After a week of consultation with the country of asylum and much scolding (not only by the embassy but by her family as well), Ladan obtained a travel document to go back.

Being extremely grateful and relieved, sometime later she nonetheless decided to travel again in order to see her husband despite the same lack of documents. We ended up in the same situation, except that this time the embassy refused to help her. They had read her previous file and concluded that she should have learned from her previous mistakes. Knocking on their door many times, an employee decided to share a story with us. The employee had once heard of a man who had been in a similar situation as Ladan, who had travelled on similar documents through the airport. He had successfully managed to fly back to the country in which he was seeking asylum. Indirectly, the employee at the embassy showed us that, despite having very clear rules on what counts as valid travel documents, in practice the system of border control was not always able to maintain such standardized rules. And so Ladan bought a flight ticket and arrived safely at her destination. In fact, the employee had obtained this *in-formation* in the same way as Fuuad, Riyaan and Ladan had gathered theirs, namely through the stories of others. He thus turned out-formation to *in-formation* for Ladan.[2]

Border work tactics

Fuuad's, Riyaan's and Ladan's decisions on where, when, how or if to move are influenced, affected and at times opposed by others who also act in what we in this book have called the biometric border world. They might be other refugees, but they are just as often those who work with and around asylum-seekers and refugees among whom the implementation of biometric technologies in the form of fingerprint registration, ID control or DNA registration have become routine, as described throughout this book. 'Routine' here refers to how work on the body has developed into an integral aspect of conducting *border work* (see Parts I, II and IV). Examples of border work are when the bus driver refuses to let Ladan on board due to an invalid ID card, or when Riyaan's, Fuuad's and Ali's fingerprints are matched in EURODAC, and they are returned to their first countries of arrival by immigration authorities and police officers. At the same time, however, we see how people working with asylum-seekers and refugees make tactical choices, here defined in the words of De Certeau as 'taking advantages of opportunities' (1984: 36–37) that arise in the moment and, as a result, how they break with established routines. This takes place when Italian border guards let potential asylum-seekers pass the border without registering their fingerprints, as in the case with Riyaan, due to their own collection of *in-formation* from the people *en route*, the media and their own experiences as Italians who see the increase of migrants on a daily basis (see Chapter 5). A European embassy employee also makes a tactical choice when sharing *in-formation* gathered about how to pass one of the most biometrically guarded places in the biometric border world – the airport – on an asylum-seeker's ID. This is a tactical choice, I argue, because this sharing of information was the only way the employee could stop Ladan from turning up again at the embassy, which he did by solving a situation, but only in the here and now and not as part of a larger routinized system. Instead, it broke with the idea of an overall strategic masterplan and was made up on the spur of the moment. This exemplifies how the representatives of the biometric border world conduct work by imposing and controlling the borders that have arisen within their own local contexts while at the same time also creating gaps and turning situations of biometrical border control into advantages of their own, notably through the use of *in-formation*. Such tactical choices are similar to those made by Fuuad, Riyaan and Ladan when they attempt to migrate across territorial borders with their biometrically registered bodies.

Expanding De Certeau's term 'tactics' by erasing his sharp contrast between the strategies of the powerful, defined as attempts at controlling and organizing spaces through a 'postulation of power' (1984: 38), and the tactics of the weak, described as 'a calculated action determined by the absence of a proper locus [. . .] the space of the other [. . .] organized by the law of a foreign power' (ibid.: 36–37), this section has shown how the flow of *in-formation* at times turns the migrant into the powerful and the system into the weak, as in Riyaan's case. Second, it has shown how the meetings among migrants, the authorities and technologies develop into something that is more than just a matter of maintaining a border crossing. The practices surrounding these borders are intertwined. Potential asylum-seekers, refugees and employees within the border world all gather *in-formation* that creates gaps that widen the room for (potential) asylum-seekers and refugees to manoeuvre and continue with their onward migration.

Physical presence and the absence of inclusion

The previous sections have explored how three Somalis *en route* search for and use *in-formation*. In addition, it has explored how they seek to avoid their bodies being turned into borders, thus restricting them in their movements, by searching for the majority of their *in-formation* from within their Somali networks. This section shows how the consequences of the biometric *out-formation* that is stored about Somali women and men *en route* led some to seek information from the system itself. By following the actions of Ali (see Chapter 5), I show how his failed attempts to seek acceptance in the EU based on his status as a refugee shows how his physical presence is tolerated in the EU while denying him the same accesses to, for example, the European job market.

A friend of mine, a Somali and a refugee himself, described the consequences of the body being a border in the following way:

> Since the technologies were implemented, the opportunities have decreased . . . The Quran prescribes how, when you die one day and you stand in front of God, on the Day of Judgement, and you will be told whether you are going to heaven or hell, you cannot lie in front of God. Your feet, your hands, your eyes, your ears have done this and that [your body will tell your story]. What should we do now?, is what many people ask themselves. Our fingers are testifying for us while we are still alive.[3]

It was these biometric testimonies that made Fuuad, Ladan and many more feel incarcerated in the countries in which they were seeking asylum. They could not erase this 'truth-teller' because it was deeply rooted in their physical existence. One young Somali man explained how he had physically tried to remove his fingerprints by heating up a flat iron in order to seek asylum in another European country than Italy. His efforts were of little avail, however, because the country to which he had fled incarcerated him for six months until the tips of his fingerprints were readable again; they then turned up as a 'hit' in EURODAC, and he was returned to Italy. The body was thus complicit in their lack of success, yet at the same time, it was that same body, its mental and physical well-being, that could take them in the right direction. Being physically fit enabled this same young man to migrate to southern Italy during the summer, where he hoped to find work on the plantations, earn a bit of money and escape life as a homeless person in the north (see e.g. Lucht 2012 for an overview of work conditions in southern Italy for undocumented migrants). Hence, the majority of Somali women and men were very aware of the *out-formation* that was stored in databases about them. Some tried to defeat or circumvent it, and a few succeeded, as this chapter has shown, but the majority failed. This led some to take a different approach, namely to comply with the rules and seek to be included in the rather complicated European arrangements for refugees, migration and work. One of them was Ali.

Ali, a Somali man in his late 20s, had realized that his biometric registration in Italy had removed his opportunities to seek asylum elsewhere. He did not want to be 'a refugee again', as he put it. This referred to his own experience as well as those of others who had tried to seek asylum in other European countries but whose fingerprints came up as *hits* in EURODAC and who were therefore deported back to Italy (see Chapter 5). His biometric registration left him with only two options, he argued: 'If I marry a woman with a European passport, preferably Scandinavian, or if I acquire a job in another European country'. Ali opted for the latter because, as he argued: 'I will try to get a job in another European country, because now I have the document'. This referred to the fact that Ali was a recognized refugee in Italy, validated in a document from Italy. He still had his Somali passport and would have to renew his Italian documents after five years. When I coincidently met Ali in Milan about a month later, he was about to leave Italy. He had found a job, facilitated by a fellow Somali who was a resident of the particular country to which he was going. Ali found the rules of the EU concerning work permits rather complicated and asked for my assistance. Diving into the European system of laws through internet searches and calls to various authorities, it became clear that Ali's status as a recognized refugee in Italy did not automatically give him the right to work in other European countries.

Ali was not a European national but a foreigner whose presence was accepted but not covered by the same legal arrangements as those that applied to European nationals. This meant that if Ali wanted to work outside of Italy in this particular EU country, he would have to find a job within certain sectors for which there was a need in this particular country. He could also find a job that was not within these particular sectors but one where he would earn a yearly salary that was significantly higher than what the average citizen in this country earned. Both requirements seemed unreachable, as the job he had secured for himself was one that did not require a long education, was not in any of these sectors and would not make him enough money to meet the salary requirement. The next time I spoke to Ali, he had found an internship and wanted me to check the law again. While Ali was searching for *in-formation* among his Somali acquaintances, I was asked to search for *in-formation* within the European legal system, maybe because I was believed to be part of it by many of the Somalis I studied and thus would be able to navigate in it. But I was not able to locate a positive answer for Ali. After that he vanished, as the majority of migrants did when they felt trapped in a world that at times would tolerate their physical presence but at the same time did not include them fully in what he saw as the unity of the EU. In many ways people such as Ali, I argue, symbolize the migrant figure, 'the political figure of movement' (Nail 2015: 11) that more often than not experiences non-inclusion shaped by national citizenship and immigration policies. They are simply too mobile to fit in!

Conclusion

This chapter has explored the ways in which biometrically registered asylum-seekers and refugees in Italy attempt to compensate for the inadequacies of the care system through the search for *in-formation*. Through the stories of Fuuad,

Riyaan, Ladan and Ali, the chapter has shown how the information they search for is first, always *in formation* and constantly changing; second, gathered among old, recent, known and unknown connections without a standardized, stringent or harmonized 'migrant infrastructure' behind them; and, third, *in-formation* is gathered in an attempt to negotiate the *out-formation* that is already stored about them in the border world they move about in. In the words of my Somali friend, they are seeking to avoid their fingers testifying for them while they are still alive, culminating in biometric registration. Sometimes they succeed when the *in-formation* they have gathered travels faster than what the European asylum system could handle, as in the case of Riyaan; or when those who were conducting *border work* turned *out-formation* into *in-formation*, as in the case of Ladan, or simply made tactical choices of their own that contradicted the task of maintaining the borders. In other cases, Somalis failed, as in the case of Ali (see also Chapter 5).

This chapter has also shown how the search for *in-formation* sometimes enables Somalis *en route* to find gaps in a biometric border world that is otherwise always being tightened through the implementation of biometric technologies, border police and EU regulations. When the *in-formation* thus gathered creates an opening, such as one positive story emerging alongside the many negative ones, as was the case with Riyaan, or when the territorial site they find themselves in becomes too insecure, they continue to migrate. Taking Denmark as an example for the latter, in September 2016 five people of Somali origin who had been accepted as refugees in Denmark had the prolonging of their residence permits rejected by the Danish Refugee Appeals Board (Flygtningenævnet). In other words, their residence permits were withdrawn because Somalia's general situation had, according to Denmark, changed for the better. This means that today Somalis can no longer obtain residence permits simply by virtue of coming from Somalia. As a result, another 1200 Somalis will have their cases reopened to see whether the same rules should apply to them (Udlændingestyrelsen 2016). This has created a small wave of onward migration by the Danish Somali community; for example, a Somali family of nine fled to Germany, where they applied for asylum again because seven out of the nine family members were to be deported from Denmark to Somalia under the new rules (Andersen 13 December 2018).

The biometric registration of fingerprints continues to play a major role in the border work of the European system for asylum, despite the fact that biometric identification does not stop attempts at onward migration when the conditions that people flee from overshadow the fear of getting caught. The latest EU initiative to manage migration influxes into Europe once again focuses on identification through registration with fingerprints, as stated in what follows:

> the primary aim would be to improve the process of distinguishing between individuals in need of international protection, and irregular migrants with no right to remain in the EU, while speeding up returns.
>
> (European Commission, 24 July 2018)

While the system keeps imposing more and more biometric measurements on people *en route*, the latter continue their search for *in-formation*, while at the same carrying around the most brutal border of them all – their bodies.

Notes

1 In fact, in Italy housing, language classes and, for some, internships are provided for the first six months after receiving refugee status. See Chapter 5.
2 See also Simonsen (2017), where parts of Ladan's story are also presented.
3 Translated from Danish to English by the author.

Epilogue

What does it mean when the tip of a finger is reduced from being part of a social human body embedded in relations and obligations of care to being a tell-tale body part that will ultimately categorize you as either an economic migrant not entitled to care or as a refugee entitled to a certain type of care? Can the scan of a fingerprint objectively determine the care you are entitled to receive and that the state is obliged to provide you with because you belong to the human species? And if so, why is this experienced as judgement day on earth by many of those who have been biometrically registered as part of the refugee care system? These questions arise from the mismatches described throughout this book between the universal human rights approach that grants a minimum of rights to every human being through international conventions, and the ascribed role of biometrics today as a universal and objective means of identifying and categorizing people *en route*.

The universal definition of the human being, as set out in international conventions, originated in the aftermath of the Second World War. It had become clear that during World War II human beings had been violated by various nation states that had destroyed the minimum conditions of life. As a way to prevent this from happening again, the international community adopted a human rights approach, as formulated in the Universal Declaration of Human Rights of 1948 and the 1951 Convention relating to the status of Refugees, amended in the 1967 Protocol,[1] that obliged nation states to provide and guarantee minimum rights to people, including those who were fleeing persecution etc. (Kathrani 2010: 116). All human beings were, at least according to the principles underlying the conventions, considered equal and entitled to a minimum of care and freedom. They were all to be treated alike and as part of one human family regardless of differences in their politics, power and socio-economic position. The EU's current implementation of the biometric registration of people *en route* follows the same idea: by acknowledging and identifying each individual, in this case biometrically, they are provided with certain rights. The argument is, in other words, that in being biometrically registered *en route,* they will be provided with the minimum conditions of care they are entitled to as economic migrants, asked to leave voluntarily or deported by force, or given refuge as asylum-seekers if they qualify under the 1951 Convention. In addition, biometrics is more often than not presented as unbiased and objective, being based on technologies that do not have the human flaws of prejudgement, emotions and so on.

This book shows, however, that nothing is ever unbiased or objective – not even international human rights conventions or biometric technologies or the combination of the two. To be able to include every single living human being, the international conventions are worded very broadly. As shown in Part IV, the 'right to a family life' is so general and abstract that it makes possible very diverse interpretations by each national state. Hence, although the international human rights conventions have guided nation states in their practices concerning refugees, there are still specific national interests that make the provision of minimum rights and freedom very diverse (Kathrani 2010). When biometrics are involved in such practices and turn out very differently both within and between nation states (Düvell and Vollmer 2011; Riedel and Schneider 2017), they only reinforce such practices of differentiation.

Nor are the biometrics themselves unbiased. They are tools without human affect and emotion, but they are constructed, developed, tested (see Part I) and implemented (see Part II) by human beings who are not and never will be objective. As also touched on in Part I, the databases used by tech developers in the laboratories to develop and test the algorithms that are used in facial recognition are biased in that they consist mostly of white males. This means that biometric technologies implemented at, for example, border sites have a higher risk of producing errors and misidentifying black people who are crossing borders. In fact, 'darker-skinned females are the most misclassified group, with errors up to 34.7%' versus 'a maximum rate for lighter-skinned males of 0.8%' (Boulamwini and Gebru 2018: 1).

These mismatches raise questions as to whether ideas of care and human compassion can be determined in a universalistic, unbiased way through the body parts of human beings. And if they cannot, what visions of the human being are created when body parts are broken into biometrically readable parts that say little about the social networks of relations that individuals are part of?

Note

1 'As a result of events occurring before 1 January 1951 and owing to a well-founded fear of being persecuted for reasons of race, religion, nationality, membership of a particular social group or political opinion, is outside the country of his nationality and is unable, or owing to such fear, is unwilling to avail himself of the protection of that country; or who, not having a nationality and being outside the country of his former habitual residence as a result of such events, is unable or, owing to such fear, is unwilling to return to it' (UNHCR).

Part IV
In the family

Karen Fog Olwig

Introducing the site

As the preceding chapters have shown, with biometric registration and the possibility to check migrants wherever they are encountered, the border has essentially become located within the body. This final part of the book will examine a form of biometric border control that is not only located within the individual body but also simultaneously shared with, and invisibly linked to, other bodies located in different countries. By focusing on refugees who have obtained asylum in Denmark, I investigate how DNA testing and X-rays that create digital images of genes, bones and teeth have become key elements in a complex series of decisions determining individuals' relations of kinship and thus their right to be reunited with family members who have been scattered as a result of flight. Like the studies of scientists developing new technology and of migrants *en route* in Parts I and III, this ethnographic analysis emphasizes that the biometric border world does not necessarily have a clear material presence in the shape of conspicuous physical sites of control, such as the ABC gates in airports or the high walls monitored by surveillance technology in Ceuta, described in Part II. Instead, it can take the form of much less visible, but nonetheless extensive systems of registration, verification and assessment that permeate, and cross the boundaries of, both human bodies and nation states.

Family reunification

Migration through family reunification has not received the same attention as the spectacular attempts to storm the Schengen border in North Africa, the hazardous crossings of the Mediterranean in flimsy boats or the dramatic movements through Europe of large groups of people in 2015, which led several Schengen countries to reinstate border controls. Family migration, however, accounts for a substantial part of actual and potential migration to Europe. Thus, according to a European Commission report, in 2015 '38% of all valid residence permits issued in the (Member) States were for the purposes of family reunification' (European Commission 2017b: 10). In the shadow of this relatively high figure, furthermore, is an unknown but considerable percentage of people who have failed to obtain the right to family reunification, largely because of the biometric testing of their relatives.

Family members often become separated in the course of their escape. Individuals frequently need to take sudden flight with little warning, making it, also, time-consuming and complicated to organize an escape for their entire family; parents may decide that it is too dangerous for their children to undertake what is often a long, hazardous journey to a distant destination, or they may simply not be able to finance the expenses involved when bringing a whole family to a safe country. For refugees, becoming reunited with family members who have become dispersed is thus an essential part of their efforts to develop a reasonably normal life in a new society that lies across many physical borders. In this endeavour, they are protected by international conventions. The UN Declaration of Human Rights of 1948, signed the same year by Denmark, states: 'No one shall be subjected to arbitrary interference with his privacy, family, home or correspondence' (United Nations 1948), while the European Convention on Human Rights (ECHR), ratified by Denmark in 1953 and incorporated into Danish law in 1992, affirms simply: 'Everyone has the right to respect for his private and family life, his home and his correspondence' (Council of Europe 2018). The documents do not explain what is meant by 'family life' but implicitly leave this key issue to the discretion of individual countries. The UN Refugee Agency encourages states to adopt a 'culturally sensitive' and 'inclusive definition' of the family and to 'apply liberal criteria in identifying those family members who can be admitted with a view to promoting a comprehensive reunification with the family' (United Nations Refugee Agency 2012). According to the Danish Immigration Appeals Board, however, the European Court of Human Rights generally bases its decisions on the view that family life is a matter of 'the nuclear unit common in European tradition' (Udlændingenævnet 2014: 104). Indeed, most European countries, including Denmark, limit refugees' reunification to the nuclear family defined as parents and their biological or legally adopted children younger than 18 years of age (Heinemann et al. 2015: 5).[1] Moreover, individuals must prove that they are genuine members of the nuclear family, either by means of legally valid documents or, if they are not in possession of any official papers, by other means. In Denmark, biometric documentation, as an alternative to legal documents, has been an important 'other means' since the 1990s.

The Danish Immigration Service

Applications for family reunification in Denmark are processed by the Immigration Service as part of its wider responsibility to handle all 'cases concerning foreigners' right to visit and stay in Denmark' (Udlændinge- og Integrationsministeriet 2018). This government agency has been reorganized, and renamed, several times in response to criticism of its administrative practices (*Politiken* 2011). The best-known instance of its mismanagement is the 'Tamil Case'. This concerned the blocking of final approval of 130 to 140 Tamil family reunification cases by the Conservative minister of justice, who was opposed to the Danish Immigration Law of 1983 that granted refugees the right to family reunification as soon as they received asylum (Landsted 2018). The case led to the fall of the government in 1993 and the restructuring and removal of the immigration agency from the Ministry of Justice to the Ministry of Social Affairs (Hertz 2017). In 2001, it was moved again, now to a special Ministry for Refugees, Migration and Integration as part of efforts to introduce tighter migration control. In 2006 it underwent yet another restructuring after severe criticism by the auditing committee of the Danish parliament (Folketinget) and was renamed the Immigration Service (*Politiken* 2011). But only two years later, in 2008, the media presented documentation showing that the Immigration Service had failed to inform people about their rights, leading the parliament's ombudsman[2] to initiate an investigation that uncovered serious problems in its practices (Folketingets Ombudsmand 2008). When some observed ironically that 'Service' might be a misnomer for a governmental agency with so many issues, the director retorted: 'That is not our experience; we have not met many who think that it is misleading that we are associated with the concept of service' (*Politiken* 2011). The Immigration Service underwent further restructurings in 2011 and 2015.

In recent years, as the Danish political landscape has become increasingly divided, the entire immigration system has become a topic of critical public debate. Nationalist groups are calling for ever tighter immigration regulations, even when they involve 'going to the "edge"' of international human rights conventions, whereas more internationally oriented groups argue for a more liberal

Figure IV.1 The Danish Immigration Service

Source: Photo by Kenneth Olwig.

immigration policy and emphasize the need to respect international conventions. These divisions are apparent in public debates on family reunification, as reflected in two stories, with opposite implications, that appeared in the media in 2016, as I was beginning my fieldwork. One of them described how the wife and children of a Syrian refugee, called Ahmed, were 'stuck' in a refugee camp in Jordan. They had no passports and were therefore unable to cross the border to the Danish embassy in Beirut, Lebanon, as required by the Danish Immigration Service in order to be DNA tested to prove that the children were Ahmed's biological offspring. The story included critical comments from several 'experts', who questioned the Immigration Service's demand that DNA testing had to be performed in a Danish embassy in Lebanon when it was well aware that the family did not have the requisite documents to enter the country. It was even claimed that this practice might be against international conventions (MetroXpress 2016). Another story concerned another Syrian, Daham Al Hasan, of whom it was reported that he would receive almost 215,000 kr (29,000 euros) in annual child support for the 17 children with whom he had been granted family reunification. Presented with this case, several members of the parliament expressed outrage and demanded that the Immigration Service DNA test the children, in case this had not already been done, to make sure that Daha Al Hasan was the biological father of all of them. This led an immigration lawyer to remind the politicians they had no business interfering in the Immigration Service's handling of cases (Sølvsten 2016). While the first story thus was critical of the rigid adherence to formal rules and regulations, the second led to complaints about leniency in the granting of family reunification. These news stories necessarily presented simplified accounts of the often complicated course of events that can unfold in connection with individuals' flight and relocation to another country. However, the different perspectives on family reunification they represented also demonstrated clearly that the Immigration Service was operating in a veritable political minefield.

I decided early in my fieldwork to ask for an interview with officials in the Immigration Service, which was playing such a key role in family reunification. It took more than two months to get an appointment, and it was therefore with great anticipation, curiosity and some nervousness that I made my way to the Immigration Service, centrally located in Copenhagen. A senior official from the Department of Family Reunification met me and took me to her office, where the interview would take place. As she guided me through the complicated corridors and stairways connecting the several buildings that over the years had been incorporated into the expanding Immigration Service – and, as later became apparent, represented an apt architectural metaphor for the complex and unwieldly bureaucracy of the Service – she said she hoped I would understand that they could only give me about an hour. They received requests for interviews from a large number of people who were interested in migration issues in Denmark, ranging from journalists to schoolchildren, and there was a limit to how much time they could spend in such meetings. They were very pressed for time, especially after the onset of the 'refugee crisis' in 2015, when thousands of people from the Middle East and Africa entered the country, overburdening the Immigration Service with applications for asylum and, later, family reunification. But, of course, she assured me, if I needed more time for the interview, they would try to accommodate me.

Worried about the interview's limited timeframe, I went straight to the core of my research interests and asked what had led the Immigration Service to adopt biometric testing as a way of documenting family relationships. Some refugees, the official explained, did not respect the Danish rule that family reunification is only allowed for members of the nuclear family but attempted to include other relatives in their applications as well. It therefore became necessary to require biometric proof of family relations if no other form of documentation was available. When I asked about some of the problems that might arise in connection with biometric testing, referring to the recent articles I had read in the Danish media, she replied that one should not trust everything that one reads in the newspapers. Contrary to the impression that one might receive from newspaper articles, she added, the Immigration Service made careful, individual assessments of each case based on the 'political decision to give family reunification to only certain groups' – the 'certain groups' being well-documented nuclear family units. And in this work, she added, biometric assessments were of key importance: 'In our procedures for processing cases, DNA [analysis] and [biometric] age [assessments] are really some of the most fundamental when it comes to determining who you are and your relationships'.

Biometric testing of the family

The biometric technologies employed in assessing cases of family reunification are well established, with DNA analysis dating back to 1985 (Heinemann et al. 2015: 1) and X-rays of bones and teeth even further back in time (see the Introduction). As is the case with other forms of biometric border control discussed in the previous chapters, DNA analysis and X-rays of bone and teeth are expected to produce objective, scientifically sound forms of identification. However, unlike the well-known digital images of faces and fingerprints used today, DNA and X-ray analyses are not carried out to record and register individuals' identity but to determine their personal family relations. This takes place through a process whereby a scientific biometric analysis determining the biogenetic ties and biological maturity of individuals is transformed from denoting, respectively, a biological procreational relationship and a biological age to signifying, respectively, a parent–child relationship and a child who is legally dependent on parental care. Within the specific contemporary context of family reunification policy, in other words, the results of the biometric tests come to be interpreted according to particular notions of what constitutes a proper family in society. In Denmark, as I show, the DNA test came to stand for a range of emotive issues tied to pre-existing discourses concerning the imagined role of the nuclear family as the nucleus of the Danish nation and the threat posed by the 'un-Danish' family relations of immigrants and refugees, as well as the large number of reunifications that they might entail.

Being curious about the processes whereby biometric tests more concretely are interpreted as evidence of family relations, towards the end of the interview I asked permission to interview other staff to learn about their experiences in handling different cases of family reunification. My request was denied. Not only did they not have the time to contribute to my research, but they also had to protect the privacy of the refugees and their families. If I wanted more details, I could consult the home page of the Immigration Service, where all the regulations and procedures were described carefully and in detail. If I needed specific further information, I was, of course, welcome to contact them again.

Fielding the site

Further research within the Danish Immigration Service could have generated very interesting data on how officials approached different cases. What kinds of questions did they raise in specific cases? What evidence did they deem necessary? And what sort of considerations led them, finally, to make particular decisions? Without this insight, I was largely restricted to examining the more problematic cases that ended with various legal services, where refugees sought help when they wished to appeal against negative decisions. This limitation on my access to the Immigration Service, however, also opened my eyes to other opportunities that led me to develop a different approach to my research. I became aware of the much wider field of geographically, politically and socially diverse institutions, actors and policies that individuals have to negotiate when seeking family reunification. Exploring this field proved to be a fruitful project generating rich data on the multifaceted role of biometric assessments in cases of family reunification.

I entered this new field as an anthropologist who has done extensive research on the role of family relations in migration processes, including the equivocal expectations of the 'integration' of families into Danish society (Olwig et al. 2012; Olwig 2015, 2018; Olwig and Paerregaard 2011). This background led me to examine the varying and often conflicting interpretations and practices of migration and integration expressed by the differently positioned actors in the field, with a particular focus on the significance of the biometrically defined nuclear family unit as a symbol of adaptation to life in Danish society.

In the present study, these actors included employees, mostly former ones, of the Immigration Service who had experienced working in different departments of this government agency; staff at Danish embassies performing mouth swabs of individuals seeking family reunification in Denmark; professionals knowledgeable about and involved in the analysis of DNA tests and biometric age assessments; volunteers and staff at formal and more informally structured NGOs offering legal and practical assistance concerning family reunification; and, finally, refugees who have been involved in applying for family reunification themselves or have close relatives or friends with such experience. I would have liked to have interviewed a broad spectrum of the members of the parliament (MPs) who, in the course of my fieldwork, passed a large number of laws restricting the rights of refugees in Denmark. The only MP who agreed to an interview, however, was from a party on the far left that represented one of the few consistently critical voices against the government's policy on family reunification. I did find nonetheless a great mine of information on political debates and law-making, as well as on legal decisions made in courts of appeal, in the extensive government documents that have been digitized and placed on the internet. Furthermore, the digital archives of different newspapers and the web pages of various NGOs were excellent sources of information on popular activist engagement. This biometric border world became my field site.

My fieldwork can be described as involving a diffuse, extended fieldsite in the sense that it examined a dispersed and fragmented assemblage of actors (see the Introduction). It also had a strong local component, however, because it involved participant observation in various localities that were not just scattered, partly connected touch-down points in a biometric assemblage, but also places in their own right that contribute to shaping the biometric border world. They include NGOs in different parts of Denmark offering legal and practical help to refugees, hospitals performing biometric testing and analysis, Danish consular offices serving as intermediate links between the Immigration Service in Denmark and individuals abroad involved in family reunification cases, and, finally, Danish society as a specific place where family reunification policies have acquired a particular meaning and form. My fieldwork can therefore more accurately be described as having taken place in a multi-local, extended fieldsite (Olwig 2007).[3] These characteristics of the fieldwork have influenced the kind of ethnographic data that were generated. Thus, rather than a number of in-depth cases examined from different perspectives and followed through time, this ethnography consists of the many different, often only partly illuminated cases, stories and incidents I encountered in different contexts, whether at NGOs assisting refugees with their applications, at consular offices performing DNA tests or in the homes of refugees relating their experiences. While it was frustrating not to be able to explore how particular cases were interpreted by different actors or to obtain information about their eventual outcomes, the partiality of knowledge, understanding and interaction within reach in this field site was, as shall be seen, very much a characteristic of this part of the border world.

The chapters

This ethnographic site is thus one of the implementation, perceptions and consequences of using biometric technologies to verify kin relations when refugees seek family reunification. The two chapters that follow show how the biometric border that controls family reunification extends far beyond the actual physical sites and temporal moments of border crossing, with far-reaching ramifications for the refugees and their families. Together they present rather different perspectives on the biometric testing of family relations. The first, more historical chapter examines the introduction during the 1990s of biometric technology in the assessment of family reunification with particular attention to refugees from Somalia, a group that, as will be seen, played an important role in precipitating Danish concerns about the need to institute biometric verification of family relations. The chapter discusses the disparate logics of family and kinship as either biologically fixed or socially generated that became apparent with biometric testing, as well as how they reflected the refugees' situations in Denmark and further shaped their experiences of settling into a new society. I show that for the Danish government, as for many Danes, biometric testing came to signify a means of safeguarding the close relationships within the nuclear family, the right to dwell in Denmark under international universal human rights conventions and the ability to maintain a proper family life suitable for Denmark's welfare society. For the Somali refugees, however, the introduction of biometric technology to distinguish biologically related from non-biologically related nuclear families had mixed effects. Those who were in the process of establishing a family life in Denmark found it created serious difficulties by narrowing reunification to the biometrically defined nuclear unit. Many, however, who had settled in Danish society came to realize that it presented them with an opportunity to renegotiate transnational family obligations that were becoming more and more difficult to honour.

The second chapter focuses on the present-day situation regarding family reunification, when ever more elaborate and complicated application procedures have been introduced. It also discusses how the many different officials who act on behalf of the 'state' become involved in testing and determining individuals' right to a family life and on what basis. It argues that, as the immigration and refugee system has become more and more extensive and elaborate, no individual person can be identified as actually taking responsibility for decisions in family reunification cases. Rather, assessment of family reunification is the result of a number of separate decisions made in disparate parts of the border world. It has essentially become the responsibility of a large bureaucracy using a seemingly neutral 'scientific' approach in which 'objective' things like the swab used to collect saliva for DNA are given central authority, thus hiding the key role of human interpretation and decision-making. This is problematic because biometric testing can have seriously deleterious consequences for the biologically related nuclear family it was meant to help. Thus, it may contribute to keeping family members separated for years, in some cases even making it impossible for them to reunite.

Notes

1 The rules are more restrictive for immigrants.
2 The ombudsman is a legal expert elected by the parliament to investigate complaints about the public administration (Folketingets Ombudsmand 2019).
3 See Wahlberg (2018) for a somewhat similar approach to the study of a social phenomenon by conducting 'assemblage ethnography'.

7 Biometric verification versus social validation of relations of kinship

Somali refugees in Denmark

In 1989, two-year-old Hanad and his twin brother arrived in Denmark from Somalia. Their father had been closely associated with the socialist political regime under Siyad Barre, and when it was collapsing, and a civil war was looming, their mother wanted to bring them to safety in another country. She therefore asked Hanad's older brother, who had already received asylum in Denmark, to include the young twins in his application for family reunification along with his wife. He obliged but decided to list them as his own children, fearing that the Danish immigration authorities would not accept the young brothers as part of his family. 'My older brother said', Hanad explained, '"I didn't want to take a chance. I would rather lie than risk that you would not be able to come"'. Hanad and his twin brother therefore came to Denmark as the children of their older brother and grew up in his home with his wife and their children, who were born in Denmark.

Hanad's family was among the earliest Somali refugees to seek refuge in Denmark. During the 1990s, several thousand Somalis entered the country as asylumseekers or as the reunited family members of Somalis who had obtained refugee status in the country. And, as was the case with Hanad, many of the reunited children were not the biological offspring of refugees in Denmark. Hanad recalled, 'I know other families that came at this time, who brought both cousins and other children along'. This inclusion of a broad range of children in family reunification applications, however, was brought to an abrupt halt in 1996 when the Danish immigration authorities initiated biometric verification in the assessment of family relations in cases where applicants were not in possession of valid documents proving such relations. Since the Danish government treated documents issued after the breakdown of the Somali government as invalid, most Somali families had to undergo biometric testing involving DNA analysis to check whether the children they claimed were the biological offspring of refugees in Denmark. Children suspected of being older than 18 years of age were also age-assessed on the basis of X-rays of bones and teeth. Using such biometric verification, in other words, the family itself was treated as a biogenetically defined unit of parents and dependent children. The nature of the family therefore also became implicitly linked to culturalist discourses emphasizing the importance of bio-genetic ties over inter-personal relations formed in the course of social life.

Focusing on Somali family reunification in Denmark, this chapter examines the introduction of biometric technology to assess family relations and the ways in which Somali refugees in Denmark experienced this and responded to it. I show that there was a built-in incongruity between the biometrically defined nuclear family stressed by the immigration authorities and the relationally defined notions and practices of kinship and family that prevailed among the Somalis, causing considerable complications for the latter. This conflict, however, cannot be understood only as a clash between different understandings of what constitutes a family and therefore who has the right to be reunited as a family. During the 1990s, family forms other than a married couple and their biological children were also common among the Danes. The nuclear family, however, still held a dominant ideological position in Danish society, being associated with a complexity of meanings related to pre-existing discourses regarding the central role of this type of family in Denmark's welfare society. It was, furthermore, a unit of family life that could be documented biometrically, thereby enabling a new form of 'scientifically' based border control that would keep family migration to a minimum. In this situation, the Somalis' flexible, broad-based notions and practices of the family came to be stigmatized as un-Danish and as a threat to society. This had important consequences for their ability to establish a new life as refugees who had received asylum in Denmark. The analysis therefore shows that the biometric border world can intrude into individuals' private lives many years after they have crossed a physical border.

This chapter draws primarily on interviews with Somali refugees who arrived in Denmark during the 1990s or who are the offspring of such refugees, as well as on documents from political debates in the Danish parliament and articles in the Danish press.

Biometric testing and the protection of culture

When Denmark introduced biometric technologies in the context of family reunification during the 1990s, it was one of the first countries to do so. Since then biometric assessments have been widely adopted, and by the mid-2010s they were being used in at least 24 countries (Tapaninen et al. 2019: 6).[1] Biometric verification of family relations has been subject to extensive debate and criticism.[2] From a legal perspective, Barata et al. (2015: 603–604) have pointed out that it is problematic because DNA analysis 'essentially functions as the gold standard to validate the authenticity of the claimed relationship in cases where documents cannot validate it'. Thus, even though the DNA analysis of biological kinship must have a probability of 95.5% or higher to be considered proof, it is, in fact, not '100% conclusive' – a fact that is easily forgotten.[3] Furthermore, as Dove (2013: 493) points out, the 'focus and over-reliance on the molecular can blind other assemblages – the non-somatic testimony comprised of documentary and oral evidence – that may be equally if not more important'. It is therefore 'no longer the state interpreting immigration documents in accordance with the context and meaning of the applicant, but rather, the state imposing its interpretation onto the applicant's molecularized, scriptured body' (Dove 2013: 472).

Figure 7.1 Samosas, Danish Christmas cake, apple juice and coffee. Somali hospitality
during an interview

Source: Photo by Betty Pedersen, 2017.

This critique is in agreement with current anthropological thinking on family
relations that emphasizes the need to distinguish between the biogenetic ties that
can be established through, for example, DNA analysis and the notions and practices of relatedness (Carsten 2000) and feelings of mutuality (Sahlins 2013) that

emerge in the course of social life and are given meaning as kinship and family life. Making the biometrically defined nuclear family the standard, fit-all unit of family life therefore imposes a substantial limitation on the right to a family life. Within the context of refuge and migration, however, the prominent role of the state in defining refugees' family relations is no surprise. As noted by Silverstein (2012: 11), 'what makes a family legitimate is always determined by the receiving community'. Such legitimacy, moreover, is not necessarily a question of being able to recognize a family within the legal system of the receiving country. Heinemann et al.'s comparative sociological-anthropological study of family reunification in several European countries thus shows that states often operate with a form of double standard that recognizes a wider spectrum of family relations among local citizens than among refugees and immigrants (Heinemann et al. 2013: 185; see also Abrams and Piacenti 2014: 631–632). This suggests that the significance of biometric testing lies in the power it gives the state to 'reduce highly heterogeneous and culturally embedded notions such as "family" to measurable relationships based on the western notion of the "nuclear family"' (Dijstelbloem and Broeders 2015: 26) and thus to restrict immigration through family reunification still further. Besides, since these restrictions are based on biometric testing, the responsibility for decisions taken in concrete cases seems to rest with 'scientific' proof rather than in any human considerations (Silverstein 2012: 11).

This critical literature on biometric testing in relation to the family reunification of refugees forms an important backdrop to the ethnography presented in this and the next chapter. However, rather than focusing more narrowly on the role of biometric technologies in the actual application process, I argue here that the exercise of biometric verification in the area of family reunification must be understood within the broader context of a border world that extends beyond actual border crossing. A border world, as discussed in the first introductory chapter, concerns more than the activities related to physically crossing a political borderline separating one country from another. As Donnan and Wilson (1999, 2010) among others, have demonstrated, bordering comprises a complex of ideologies, legal structures, policies, public imaginaries and social and economic interests connected to cross-border movements and relations. An important aspect of such border practices thus pertains to the reception and incorporation of immigrants and refugees in a new society.

The arrival of Somali refugees in Denmark generated anxieties about the perceived foreign ways of people from a distant African country, which sparked debates on the difficulties of 'integrating' this population into Danish society (Rytter 2018: 6; Hervik 2011; cf. Grillo 2003). It has been argued that an important aspect of this anxiety has revolved around the concern to maintain the special Danish culture that is believed to constitute the basis of Denmark's modern welfare society (Olwig and Paerregaard 2011; Schmidt 2011). Danish culture (or Danishness) is rarely clearly explicated. Rather, the positive qualities of the Danish culture of the majority population are demarcated from the 'un-Danish' cultural practices of immigrants and refugees by pointing out the latter's negative qualities. In this socio-cultural boundary work (cf. Barth 1969), I would argue, family relations have come to signify a particularly distinguishing mark of the fundamentally un-Danish, and thereby problematic, ways of non-Danes, or the 'not-yet-real' Danes (cf. Rytter 2011; Schmidt 2011). This, in turn, has provided a powerful justification for the necessity of introducing biometric verification of the family relations involved in cases of reunification. The biometrification of the Danish border world is thus part of a broader process involving increasing legal restrictions on immigration and integration introduced as part of efforts to 'protect' Danish society against 'undesirable' cultural elements that are regarded as irreconcilable with Danish society. Denmark, however, is not unique in justifying the introduction of biogenetically based, restrictive policies towards refugees and immigrants by pointing to the perceived need to protect Danish culture. The use of culturalist arguments to control the immigration of people of certain ethnic and racial backgrounds has been described as part of more general developments in Europe in the direction of 'neonationalism' and 'neoracism' (Hervik 2011).

Family reunification in Denmark and Somali relations of family and kinship

When Hanad's brother applied to become reunited with his two young brothers, he was well aware that they did not qualify for reunification under Danish law. Indeed, it could be questioned whether one could even speak about a reunification with close relatives, since the elder brother had worked in Saudi Arabia for several years and had never lived in the same home as his younger twin brothers. Nevertheless, within the Somali scheme of things, it made good sense to regard the brothers as part of his family life.[4] In East African societies, family life generally unfolds within the framework of larger kin groups. Somali family life is shaped by agnatic lineages, as well as crosscutting networks of affinal relations, and it takes place within a variety of different local entities, ranging from the monogamous nuclear family to the polygynous family and various configurations of the extended family. Siblings and cousins are regarded as belonging to the same relational category of relatives (Besteman 2016: 94), and the domestic units are characterized by great flexibility in the form of the relocation and regrouping of relatives in response to the availability of resources and the incorporation of relatives in need (especially orphaned children). Thus, Somalis are embedded in a system of mutual obligation to extend various forms of help and assistance within wide circles of relatives (see e.g. Lewis 1994; Horst 2002; Johnsdotter 2015; Besteman 2016). The 'ethic of reciprocity and of mutuality' is particularly strong within the clan, with the better off having a strong moral responsibility to help the less fortunate (Lewis 1994: 128).

Such malleable, yet dependable, relations of family and kinship comprise a key framework of social, economic and emotional welfare among Somalis. Practices of care and support are thus not necessarily concentrated on parent–child relations, as in the nuclear family, but may be extended to other intergenerational relations. Many children, for example, live with other relatives, at least for some periods, whether because their parents for various reasons are not able to take care of them or because the children can gain access to better social and economic opportunities by living outside the parental home. Children may also live with other relatives because they themselves can provide care to others – for example, younger children or elderly relatives such as grandparents – or just help in various ways. They are incorporated into these family groups and expect to be received and treated as children of the home. The flexibility of parent–child relations is reflected in the term of address children use toward the maternal aunt, *habaryar*, which means 'little mother', referring to the caretaker role that this relative would often assume.

The central importance of family and kinship in the provision of care and support has been seen as integrally related to the structure of Somali society, where the system of segmentary clans and cross-cutting affinal relations has constituted the traditional backbone of the social, economic and political order. The Republic of Somalia won independence in 1960 but was known to be corrupt and entirely dependent on aid (Lindley 2010: 25), and in 1969, General Mohamed Siyad Barre assumed power through a coup d'état. Supported by military, economic and technical aid from the Soviet Union, he promoted a program of 'scientific socialism' involving central economic planning. He had some success in increasing literacy, ameliorating the position of women and curbing 'the power of clan and religious leaders' (Lindley 2010: 25). But when Soviet support was withdrawn the regime broke down, and a civil war erupted involving opposing political factions affiliated to different clans and resulting in the flight of hundreds of thousands of Somalis (see Part III).

With the breakdown of the formal political structure, the rapidly developing conflicts and the ensuing chaos of sudden flight, the flexible system of help and support came to play a key role. Rasheed, who arrived as a refugee in Denmark in the mid-1990s, explained:

> When there is a civil war, people flee in all directions. You flee one way, another family flees another way [. . .] and when you flee your way, suddenly you have two or three children who are not yours but for whom you become responsible. Some children flee on their own, and there were thousands of children in the refugee camp with no father, no mother. They had to manage on their own. In a civil war many things happen. If the parents are not from the same clan, and the clans are in conflict with each other, the mother may not be able to flee with the children [who belong to the father's clan], fearing that she will be slaughtered. If there is no father, she may have to give the children to another person. The children may then be in another family and stay there.

The kin-based system of support not only constituted a safety net for many individuals in the course of flight; it also provided the resources individual Somalis needed to travel onwards with the aim of establishing a new family base in Europe or North America under better social and economic conditions. Somali relatives would often pool their funds to underwrite the considerable expense of the long journey, and they helped make contact with an 'agent' who could organize the complicated trip and procure the necessary travel documents, usually by purchasing, or 'leasing', a passport that would allow border-crossing into and within Europe and North America (see Part III). This assistance, as Jagd notes (2007: 249), meant that refugees in Denmark had an even stronger obligation to help their relatives left behind in refugee camps.

Most of the Somali refugees who arrived in Denmark applied for family reunification as soon as they had obtained asylum. Many, such as Hanad's older brother, chose to base the application on their own interpretation of what constituted a family. It was common to include, for example, children that the refugees had cared for before they left for Europe or simply children they knew were in particular need or wanted to help escape from the hopeless conditions of the refugee camps.[5] Most realized that this was not in accordance with the intentions of Denmark's laws, but they thought they were engaged in a higher cause – the saving of lives. As Hanad explained, 'Many children died during the war because of a lack of water or polluted water, starvation or malnourishment'. The Somali refugees often had no birth certificates proving the parentage of their children, either because they had lost them or because such documents were not usual in many parts of Somalia. Nevertheless, they had to acquire identification papers for family reunification purposes, since this was required by the Danish authorities. It was therefore common to somehow procure documents that fit the Danish requirement that the children must be the biological issue of the refugee in Denmark and younger than 18 years of age. However, the inclusion of several extra children could be complicated in terms of fitting them all into a credible chronological sequence of siblings, especially if the family was already rather large. In that case, explained Rasheed, another Somali refugee who arrived during the 1990s, the children's ages had to be 'corrected' a bit to make them all fit into a single group of siblings younger than 18 years of age. By the end of the 1990s, approximately 10,000 Somalis had acquired official refugee status, while 3,000 children had been granted family reunification (Kleist 2007: 105, 125).

Somali refugees in Denmark were not exceptional in their attempts to help as many children as possible to a safer and more stable life abroad by incorporating them into family units that would be acceptable to the receiving country. Caroline Besteman's study of the 'Bantu Somalis' shows, for example, how they had to be relocated from a refugee camp in Kenya to the US in nuclear family units. They therefore 'figured out ways to turn extended families into nuclear families while trying to avoid having to lop off those family members who could not fit into the new model', essentially by 'turning nieces, nephews, and orphans into one's children'. At the same time, ironically, some also turned their own children into orphans when they realized that adopting an orphan into the family might improve the family's chances of relocation (Besteman 2016: 93–94; see also Johnsdotter 2015 for a study of Somali families in Sweden).

The Danish border world and the introduction of biometric testing

Until the early 1990s, most of the refugees who arrived in Denmark had fled from communist Eastern Europe or Vietnam and were welcomed as persons who had resisted totalitarian regimes. When a growing number of refugees began to come from the Middle East, Sri Lanka and Somalia, they were met with increasing scepticism (Rytter 2018: 6). The Danish scepticism of Somalis, as already noted, was rooted to a great extent in perceptions that they were very different culturally and did not fit into Denmark's welfare society. In the city of Odense, for example, where about 400 Somalis had settled by 1995, the municipality began to organize 'culture courses' to help 'professional practitioners' handle the Somali refugees. A course on 'Somali family culture' drew no fewer than 68 participants (Skak 1998: 98, 105). The 'problematic' culture of immigrants and refugees was also the subject a series of articles that the Danish tabloid *Ekstra Bladet* published in the spring of 1997 (Ejrnæs 2001: 7–8; Hervik 2011). The most influential article concerned Ali, a Somali who had obtained asylum in Denmark in 1992 and was granted family reunification two years later. The article asserted that he received 631,724 DKK (approximately €85,000) in welfare for his two wives and 10 children, the message being that he had brought a large, polygynous family to Denmark, even though he was unable to support them himself. When politicians were confronted with the story by the tabloid, they stated that welfare of this magnitude to refugees was entirely unacceptable. The minister of social affairs, who was responsible for the integration of refugees in Denmark, was also reported to express bewilderment about 'what to do with a family that fits so poorly into Danish conditions' (Ejrnæs 2001: 8). Meanwhile, the government had reported to the parliament that two thirds of the DNA analyses performed in 1996 did not confirm claimed parent-child relations and that the applicants – technically the children abroad – were therefore 'fake' (Folketinget 1996–97a). This led the parliament to pass a law that made biometric testing a standard procedure in the assessment of individuals' applications for family reunification 'if their family attachment cannot be documented satisfactorily in any other way' (Folketinget 1996–97b). As a consequence, the number of Somali children approved for family reunification was reduced by half, from more than 1500 children in 1996 to 766 children in 1997, when 731 claimed children were rejected either because they failed to appear for biometric testing or because they were shown to lack the required biological kin tie to refugees in Denmark. Furthermore, 105 Somali refugees who were found to have given 'false information' in their applications lost their permanent residence permits in Denmark, and several others were brought to court and sentenced to pay for the expense of having the biometric tests performed on them (Folketinget 1997–98).

Some non-governmental organizations, such as the Danish Red Cross and the Danish Refugee Council, which work closely with refugees, expressed strong criticism of the biometric verification of family relations during a public hearing on the law. They pointed out, among other things, the problem that the law was based on a specific understanding of the family which was not universally shared and that even in Denmark the notion of the family as a biological unit would be contrary to general kinship practices (Folketinget 1996–97b). These responses did not change the view expressed by many members of the parliament that DNA testing of family relations had to be carried out to combat what were regarded as fraudulent family claims. This perception of Somali cases of family reunification seems to have been widespread in Danish society. Thus, in 2004, seven years after biometric testing had become a standard procedure in the processing of applications with no acceptable documentation of family relations, a newspaper article stated matter-of-factly that the law had been introduced because 'many Somali applications, especially, tried to get their friends to Denmark by claiming that they were part of the family'. This view was apparently shared by the Immigration Service: an immigration official was quoted as saying, '"We had the impression that many were cheating. Now the tests have had a preventative effect. The number of tests has been reduced, as has the extent of cheating"' (Bindslev 2004).

Biometric testing clearly came as a surprise to the Somalis. A close relative of Hanad's was one of the Somalis who was caught when he attempted to help his sister and her children to Denmark by claiming they were his own wife and children. After the application had been filed, they were all unexpectedly asked to have DNA tests performed – the sister and her children at a Danish embassy in a country neighbouring Somalia – and all the tests were analysed in Denmark. Several months later, he was informed that the person he had claimed as his wife was his sister and that he was not the father of the children. 'They are still down there', Hanad added.

While it was embarrassing for the Somalis to be branded as cheats, they did not accept that they had engaged in improper or dishonourable practices. Bashiir and Rasheed, who for many years had been involved in community work among Somalis in Denmark, readily acknowledged that many of the children who received reunification prior to 1997 were not the biological offspring of refugees in Denmark. They were, however, children, often relatives, in need of help towards whom Somalis in Denmark felt they had a moral obligation. Thus, whereas Somalis were fully aware that only the biometrically defined nuclear family qualified for reunification in Denmark, they did not accept the accompanying accusation that their family relations were not real or that the children were not theirs. Rasheed elaborated on this by describing a concrete case of family reunification that took place before the introduction of DNA testing:

> I know a man in Denmark who wanted to help his sister and her two children because they were born out of wedlock, there was no husband, and he said: 'they are my children'. He brought them to Denmark and onward to the US. [. . .] They are adult now, doing really well and helping their mother. And if you ask that man today, 'Did you lie to the Danish authorities?' he would say, 'No I did not. I did not lie. They are my children. I did not lie to the Danish authorities'. But if you ask the Danish authorities, they would say, 'He has lied, they are not his children'.

Somalis viewed the biometric restrictions on family reunification as being part of a smear campaign and of the general discrimination against them. This not only made them feel unwelcome in Denmark; it also exacerbated the difficulties they experienced in obtaining gainful employment. In January 2004, for example, only 23% of Somali men and 10% of Somali women between 16 and 64 years of years were employed. The remainder were enrolled in various programs intended to further their qualifications for entering the Danish labour market, such as language courses or internships, or they were on various kinds of welfare, such as the 'introductory benefits' given to refugees during their early period of asylum in Denmark (Jagd 2007: 69). Discouraged by the difficulties of settling in Denmark, a large number of Somalis who had acquired Danish citizenship chose to leave Denmark during the early 2000s and settle as EU citizens in Britain. One unofficial estimate put the number of Somalis who relocated from Denmark in 2003 as high as 1000 (Nielsen 2004: 8–10; see also Jagd 2007: 221). One of those who moved to England was Hanad's elder brother. Hanad explained:

> Even though my brother had worked in engineering in the Arab countries, and knew a lot about these machines, they kept demanding that he had to learn Danish. All he wanted to do was get a job. He said, 'I am forty years old, I can't learn Danish, I can work and learn Danish on the job', but he wasn't allowed to do that. The municipality kept sending him to different courses. So when we turned eighteen years, he thought he had done his duty towards my father, we were legal adults. And then my father gave him his freedom and said, 'Now you can travel wherever you want'.

The family biometrified

To a certain extent, Somali and Danish perceptions and practices of care can be understood in relation to two different family systems: Somali kin relations organized in segmentary patrilineal clans cross-cut by networks of affinal relations in which flexible, changeable and mobile family units are embedded versus the ideally autonomous, self-sustaining Danish nuclear family. However, the two states of Somalia and Denmark are perhaps even more radically different: the weak, or rather collapsed, Somali state where relations of care, support and security are left entirely to kin groups and networks of family relations (see Part III) versus the strong, well-organized Danish state where the nuclear family is seen as playing a central role in the social welfare system. The social significance of the nuclear family in Denmark is apparent in a 1996 paper, 'Family Obligations in Denmark', published by the Danish National Institute of Social Research, which notes that the Danish Social Assistance Act specifies that all adults have an obligation 'towards the public' to support themselves, their spouses and children younger than 18 (Koch-Nielsen 1996: 14). A large number of Danes do not live in nuclear families, however, and many of the adult men and women who do live in such families are not able to provide the kind of support that will make the family unit self-sustaining. The state therefore steps in and offers additional support to ensure the family's welfare. Indeed, a range of services are given to family units, such as partially (in some cases totally) funded day care for children, free health care, domestic services for the disabled, financial support to the economically disadvantaged and other essentials for the day-to-day care and well-being of family members. In Denmark, the nuclear unit therefore figures not only as a framework of family life but also as a key unit in the distribution of social and economic support within Denmark's welfare society.

From a Danish perspective, it is unacceptable, indeed deceitful, to create a link between the nuclear family-based system of distribution within the bounded national welfare system and the transnational system of care and support based on the broad-based system of social and economic exchange inherent in Somali relations of family and kinship. This is because making such links invariably means that Danish resources given to support nuclear units in Denmark's welfare society will be tapped, so to speak, in order to redistribute them within the wider network of relatives abroad. It was therefore thought to be completely justified – and morally right – to require the biometric verification of family relations, despite the fact that it reduced the family to a biogenetic entity and ignored the significance of social relations. From a Somali perspective, on the other hand, the biometrically defined nuclear unit represented a narrow interpretation of family relations that did not recognize the importance of their tradition of fostering orphans and other children in need. For them, moreover, fostering was not only a matter of providing vital care; it also created a strong sense of relatedness between the caregivers and the fostered child that they often interpreted as a parent–child relationship. As expressed forcefully by the Somali who helped his sister and her two sons get to the US, 'I *am* their father'.

When Somalis realized that their applications for family reunification would be verified biometrically, some sought ways to circumvent the system. One common avenue was to bring in family members on fake Danish passports to Europe or North America, where they would ask for asylum. Rasheed recalled:

> There was a time when it was possible to buy Danish passports here in the city. You go to a shop, buy a Danish passport, picture and everything, and you 'are' a Danish citizen, take the plane and travel to the US. Most of those who went to the US did it that way. You had to pay 5,000 kroner [approximately €670] for a Danish passport. And once they arrived in the US they went through the passport control and asked for asylum.

Travelling on a fake passport, produced in a small local business, became much harder when biometric fingerprints were introduced in Danish passports in 2012 while seeking asylum in other European countries became difficult, if not impossible, as demonstrated in the discussion of Somalis *en route* (Part III), after the establishment in 2003 of, EURODAC, a biometric database that stores the fingerprints of individuals who have applied for asylum in Europe during the past 10 years.

At the same time as the Somalis regretted that it was becoming impossible to help children in need through family reunification, many began to realize that practising Somali care relations in Denmark was not as unproblematic as they had thought, even if they were living in a seemingly well-resourced society. As already noted, most of the Somalis in the first generation of refugees in Denmark were not able to obtain gainful employment, and there were many expenses connected with raising the children. As Bashiir noted, 'We thought that this was paradise and that if you come here you can earn money'. Things became even more complicated when the fostered children had parents in African refugee camps who expected to be included in the benefits received from Denmark's welfare society. Child support (*børnepenge*), given to all parents regardless of income, created special problems. Rasheed related:

> I know some families [in Africa] that called the family here and said, 'I have heard that you receive lots of money because of our children. You have to send us money'. And the family here says, 'the money that we get is to support your children here in Denmark. We cannot send you any money. [. . .] If we send money, your children have no means of living. What can we do?' Then the conflict begins. [. . .] I know a couple of cases where the parents here travelled back to Africa [to return the foster children to their biological parents] and said: 'They are your children'!

In some families, tensions also developed because the fostered children thought that they were being treated harshly by the foster parents. Hanad, for example, recalled a childhood of stern upbringing that included a strict 10 o'clock curfew and being forced to go to Koran school. Only later did he realize that his father back in Somalia had told his older brother that it was his responsibility to make sure they did not become 'too Danish'. Differences in Danish and Somali views on a proper upbringing were a major source of conflict in many families. As the children learned in the Danish schools and youth clubs that they were expected to develop their own interests and become increasingly independent of their parents, the latter lost the authority over them that they would have had in Somalia. The children protested against their strict upbringing, which in some cases included corporal punishment, which became illegal in Denmark in 1997 (Skytte 2007: 78). Hanad recalled that three fostered children in a family that were close friends with his own family were beaten severely by the father. When the oldest son was hit by the father while the two families were walking together in the town where he lived, he was shocked and asked his older brother to intervene, but the latter told him not to interfere.

Intergenerational conflicts often emerge within immigrant families, although they seemed to be more severe in Somali families with foster children. Not only did these children often resent being treated more harshly than the biological children, but many also lost respect for the parents when it became apparent they were not regarded as their 'real' parents in Denmark; especially those who had joined a family in Denmark as older children and who felt they had only second-class status in the home refused to accept their place in the family. A common problem concerned older girls, who, as was customary in Somali families, had been brought to Denmark to help out in the home. As Bashiir explained:

> In Somalia there are some girls who live in the home and help out, but here you have to manage everything – you shop, make food, send your children to kindergarten, work – life here becomes difficult. And then some say, 'Oh, I will bring my sister's daughter, she can come here and help me. Then she will perhaps get an education, but in exchange she has relieved me'. Things may work out really well during the first year, perhaps the second year, but then the pressure becomes too much, and the girl says, 'I am not afraid any more. I am doing everything, it is not something we do together. I am the only one cooking every day, I am the one changing diapers'. And then she flees.

Fleeing meant reporting the parents to the Danish authorities and asking to be moved to a youth home under the care of the municipal authorities. Such 'flight' from home was feared by the parents, who felt it brought shame on the family in the Somali community and further stigmatized Somalis in Danish society.

A biometric blessing?

In light of these difficulties in practising Somali family relations in Denmark, many began to view the introduction of DNA testing as a blessing because it could simplify family relations. Bashiir was one of them:

> I think it is a great relief. Many Somali families that come to Denmark would never, never dream of bringing any [extra] children, because the longer you live in Denmark, the more you say, 'I left that [family system]. I can help my family in other ways. Perhaps I will send $50 every month, or $100 or whatever'. The more you stay here, the more your mind is opened and you say, 'Yes, I am part of the society, and if I do this, it is a crime. This is not right, and it will give me a criminal record'.

Rasheed was also in favour of DNA testing because it made it absolutely clear that it was not possible to bring children who did not fulfil the Danish criteria for family reunification.

Acceptance of DNA testing was even more pronounced among the young Somalis who had experienced the conflicts that could arise when practising Somali family relations in Danish society. They had seen some foster children treated badly by their parents and were aware that this was often reported to the local authorities and had given the Somalis a bad reputation. Hanad was both outraged and embarrassed when he witnessed his friend being hit by his foster father on the street in full public view. The young Somalis also found it problematic that the fostering of Somali children in Denmark through illegal family reunification resulted in many children being uncertain of their family relations.

Some children, for example, were not aware that they had grown up with foster parents, and several young Somalis mentioned the case of a young man who was shocked when an older Somali informed him that he belonged to another clan than that of his supposed family. When he asked his mother about it, she admitted that she was actually his mother's sister, which meant that he was part of his biological mother's husband's clan.[6] Hanad himself remembered being somewhat confused about his family relations. While he was fully aware that his mother was in Africa and he had regular contact with her, it was not until he was 11 or 12 years old that he realized he had a father in Somalia and was being raised by his elder brother.

The young Somalis disliked this uncertainty and confusion. As Hussein said, 'it is unfair not to know who your parents are'. This statement, however, reflects a certain disagreement within the Somali community about what makes relations of family and kinship. Young Somalis who had grown up in Denmark were influenced by Danish notions and practices of family relations. They therefore rather consistently used the term *real parents* when referring to their biological parents, thus stressing the primacy of blood relations, as is common in Denmark. And while they emphasized the importance of the wider support of family relations, they did not want these relatives to interfere too much in their lives. Najma, for example, thought it was wonderful that her maternal aunt in England had participated over the phone when she gave birth to her son, but she did not appreciate her aunt's admonishments about how to feed the baby. The young Somalis I interviewed wanted to live in their own nuclear families, have just a few children and be part of middle-class Danish society. As Hussein explained,

> It is traditional to have many children, right? But this is not possible here because people who have to care about their studies and want a job in the future, they cannot look after seven children. At least I don't think so. And I have enough to do in relation to my family here. I don't have such close ties to those in Africa.

While they could understand their parents' desire to help relatives in need, for them such help had become a more abstract, general question of helping people in need, rather than a personal issue of being responsible for the well-being of family members. When those who needed help during the mid-2010s were mainly refugees from the Middle East, the young Somalis seemed to have accepted a common Danish view that there is a limit to what Denmark can do. As Najma explained, 'If it was possible I think that everybody in the whole world should have a good life, but what can you do? I understand it is difficult for little Denmark to receive so many people'.

Conclusion

The Somalis who arrived in Denmark during the late 1980s and early 1990s met a relatively open country where their requests for asylum, and later for family reunification, were approved rather easily. This made it fairly simple for them to include various children in need in their applications for family reunification and thus honour the obligations they felt toward the families they had left behind under difficult circumstances. However, the generally welcoming attitude that the early refugees enjoyed changed in the course of the 1990s, as Somali family reunification practices started being condemned as a misuse of the rights extended to them as refugees, as exploitation of the Danish welfare system and as a reflection of alien cultural practices that made them unsuitable for life in Danish society. While the reaction to Somalis was especially hostile, it was part of the more general, increasingly negative response to immigrants and refugees that was emerging at this time. It was, as Rytter noted (2018: 6), closely linked to a growing Danish concern about the problem of integrating immigrants and refugees from countries with particular religious, ethnic or racial backgrounds into Danish society. The increasingly restrictive policies concerning the granting of asylum, family reunification and welfare can therefore be viewed as part of 'an overall national political strategy of social engineering with the purpose of altering the family life, religion and traditions of immigrants in order to make them become Danish' (Rytter 2018: 6). DNA testing was a key element in this 'social engineering'. It can thus be argued that this biometric technology was introduced in order to avoid 'interpreting immigration documents in accordance with the context and meaning of the applicant' (cf. Dove 2013: 472). By reducing individuals to 'molecularized, scriptured bodies', the Danish state essentially cleared the way for instituting among Somalis the bounded, biometrically defined nuclear units that were regarded as the appropriate framework for family life in Danish society. In this way, DNA analysis created a rupture in Somali kin networks that, in effect, came to constitute a border barring many from joining their relatives in Denmark.

The introduction of DNA analysis to assess undocumented applications for family reunification was not generally viewed in Danish society as a restriction of refugees' rights. It was regarded as a natural consequence of the ideology, validated by law in many European countries, that family life 'naturally' unfolds within nuclear units because blood ties constitute the 'real' parent-child relations. In Denmark, moreover, this ideology became closely linked to the perceived need to prevent outsiders from exploiting the resources of the Danish welfare state created by generations of supposedly hard-working Danes (cf. Jöhncke 2011). From this perspective, biometric testing came to be regarded as a way of protecting, or caring for, Danish culture and Danish society (cf. Chapter 5).

The ideology of the family as the foundation of Danish family life, as indicated earlier, did not reflect the 'realities' of Danish family life during the 1990s. At the time that the biometrically defined nuclear family was established as the only acceptable model for refugees' family life in Denmark, Danish family relations had undergone a long period of change, and a variety of family units now constituted the framework of everyday family life. In fact, in 1991, at the same time that Somalis were beginning to apply for family reunification, Statistics Denmark recognized the need to specify different types of family units in order to carry out more comprehensive statistical analyses of Danish families (Mehlsen et al. 2015). Several variants were added during the following years, and by 2014 Statistics

Denmark had established 37 different family types for the purposes of statistical analysis (Bindslev 2004). They included a host of different combinations of a couple (of one or two genders) with or without common biological children and possibly various children from the couple's previous relationships living in the family unit on a part-time or full-time basis (Restrup 2014). All of them, in other words, can be regarded as variations of the nuclear entity.

This official recognition of different versions of family life did not lead to changes in assessing family reunification.[7] On the contrary, during my interview at the Danish Immigration Service in 2016, I was informed that biometric assessments were necessary due to cultural clashes in understandings of the family and that they had an important preventive function. In other words, by insisting on the need to impose a biometric definition of family eligibility, the Danish authorities created a schism between the 'Danish' nuclear family unit, which was seen to be well adapted to life in Denmark, and the 'un-Danish' family relations of refugees, which were presumably ill suited to life in Denmark. While the Somalis were incorporated into Danish society in the sense that they were given asylum and the legal right to family reunification because of Denmark's obligation to honour the EHRC, they were generally perceived as 'unintegratable', having become 'the dominant Other of the Danish discourses of neonationalism and populism' (Hervik 2011: 68), and they were therefore rejected as fellow citizens. They were, along with many other refugees, placed in a permanent position of what can be called 'embedded detachment' in the Danish border world, people whose legally valid right to receive asylum in the country had to be honoured according to international conventions on universal human rights but who had not earned a place as socially accepted members of Danish society. In this border world, firm legal checks on refugees' claims to a family life were regarded as necessary and legitimate, even if this meant reducing family relations to a matter of biogenetic relations.

The Somalis' reaction to biometric testing was characterized most of all by feelings of ambivalence, reflective of their position of embedded detachment in the Danish border world and the growing tensions they were experiencing within their families. They did not like seeing their family life reduced to a biometrically defined nuclear unit and initially displayed great creativity and energy in attempting to outwit the Danish immigration system. However, as they settled and began to experience the difficulties involved in maintaining a Somali form of family life in Danish society, many, especially in the younger generation, came to accept the biometrically defined nuclear family as a legally mandated social unit that could help them renegotiate their transnational family relations. The biometric testing thus made it possible for them to limit their transnational family obligations and focus more on their family relations within Denmark. Some of the younger Somalis even emphasized that for them, the nuclear family unit was an important ideal that they had come to value. Such statements could be viewed as merely an attempt to win acceptance by professing to share the social norms and cultural values of the dominant majority population, a strategy that has been described in other studies of refugees and immigrants in Denmark (see also Rytter 2018: 11). This interpretation, however, may be too narrow in this case. Given that notions of relatedness and feelings of mutuality, which are viewed as central to family and kinship, are continuously constructed and reconstructed in light of the routines and challenges of everyday life, one would expect the Somali family to undergo changes in Denmark both as an ideal and in terms of practice. In this situation, the introduction of biometric testing became both a technology of restriction and control and a possible step towards adaptation.

However, Somalis' ambivalent acceptance of biometrics should not be regarded as an option they actively pursued. The acceptance of biometrics was rather a result of their wish to find a way out of their protracted social exclusion in a border world where they had lost the right to determine their most intimate relations because their right to a family life had become a matter of biometric verification rather than social validation by the individuals concerned. While many younger Somalis have adopted the nuclear family form, this does not mean that they have enjoyed full acceptance in Danish society, where they tend to be identified primarily as members of an ethnic minority.

With biometric identification and verification becoming more and more prominent in border worlds throughout the world, the redefinition of the socio-cultural unit of the family as a biogenetic entity has important implications for refugees' rights and inclusion as fellow citizens in the societies of refuge. It means not only that their ability to establish a family life of their own in a new society is highly circumscribed but also that, in effect, they are forced to abandon informally adopted children who are not part of the biometrically defined unit, even if these children may have lived with them for many years, are dependent on them, and have become an integral part of their family life.

Notes

1 Britain began DNA testing in assessing family reunification applications during the 1980s as the first country in the world to do so (Heinemann et al. 2015: 1).
2 For a recent review of the debate, see Tapaninen et al. (2019)
3 Barata et al. (2015: 622–623) encountered a case where retesting showed the first DNA test to be erroneous. They add that 'parentage testing laboratories in Europe reported an error rate of 0.08% in calculating paternity' in 2008, caused by 'errors in genotyping, clerical and nomenclature errors' (2015: 631–632).
4 This discussion draws on Olwig in press.
5 See Besteman (2016: 93–94) for similar practices when the 'Somali Bantus' were relocated in the US.
6 Whereas clan membership appears to be unaffected by fostering, parenthood can be defined both biologically and socially. See also Johnsdotter (2015: 86).
7 Heinemann and Lemke (2014: 491) make a similar point, arguing that DNA verification of family reunification cases in Germany established 'a double standard of family recognition for native citizens and immigrants. While for the former a variety of family arrangements is legally acknowledged, the latter have to comply with a narrow and biological idea of the family'. See also Heinemann et al. (2013: 185).

8 Mouth swabs and other techniques of verification

Determining refugees' rights to a family life

The consular officers at the Danish embassy are getting ready to perform mouth swabs for the DNA testing of individuals applying for family reunification with refugees in Denmark. Five families have gathered in the little yard by the office where the tests will be administered by the consular officers, who, between them, speak the three languages most common among refugees in this area of Africa. The officer who will do the actual mouth swabbing whereby buccal material is harvested for the DNA analysis has entered the yard where the families are waiting. The other officer, who will register each individual tested, hand out the equipment used for the testing and receive the test materials, stays in the office by a window, allowing her to follow the procedures. The window has a small opening below the glass through which she passes, to her colleague outside, the sample forms to be signed by each person, the swabs used to collect the saliva, the small cards on which the saliva is rubbed and the lollipops that are given as a reward to the children. She will receive back from her colleague the signed forms and completed tests. The consular officers have already inspected the materials brought by the families: four photographs and the 'Letter from Administration for Refugee and Returnee' for each person to be tested, the letter being the official document that identifies individuals and allows them to travel locally.

The consular officer in the yard puts on plastic gloves and calls the first family to be tested, a mother and three young children. The mother will have the mouth swab first, I am told, so that the children can see it does not hurt. The consular officer checks that the mother is the same person as the one on the four photographs and asks her to sign a form certifying that she is the person listed in the application for family reunification. The mother signs, knowing that this is required, even though she cannot read the form because it is in English. The consular officer hands the mother a glass of water and asks her to wet her mouth. Then she rubs the inner part of the mother's cheeks with the foam rubber end of a swab. She presses the liquid out of the foam onto a demarcated circle on the pink part of the card and makes sure that it turns white. When she has passed the card with the sample, the signed sample form and the mother's identity document to her colleague in the office, she turns her attention to the children, first the two boys according to age and then the girl. The children carefully sign their names and go through the same testing. When they are all done, the children receive their lollipop reward. The next family is then called in.

Figure 8.1 DNA testing of a child

Source: Photo by Karen Fog Olwig.

> *During the testing the consular officer in the office behind the window has been busy registering the names of those who have been tested, passing the required papers and equipment through the hole below the window and interpreting for the officer outside, who does not know the language spoken by the families being tested that day. As soon as a family's tests have been completed, she makes a copy of all the documents and returns the personal identification papers to them. When all the families have been tested, the consular officers comment that it was a good day – none of the children became upset and cried when they had to have their mouths swabbed.*

> *The following day the officers put the tests into five envelopes, one for each family, and place them, along with the list of all the tested individuals, in one package that is sent to the Department of Forensic Medicine at the University of Copenhagen. Here DNA analysis of the tests will be carried out, and the results sent to the Department of Family Reunification in the Danish Immigration Service.*

DNA testing of relatives included in applications for family reunification in Denmark is always carried out at a Danish embassy to ensure that it is performed properly and that the individuals tested are indeed those mentioned in the applications. The care with which individuals are checked and the buccal swabs are performed demonstrates the importance attached to maintaining the status of DNA testing as the 'gold standard' in the verification of family relations. This could give the impression that the assessment of cases of family reunification has just become a matter of performing meticulous buccal swabs and scientifically accurate analyses of the collected material. Indeed, as noted in the previous chapter, the use of biometric testing to identify family relations has been subject to scholarly critique for reducing individuals to biogenetic matter and ignoring relational family identities that are constructed and validated in social practice and narrative. Biometric testing, however, does not stand alone when Danish immigration officials ascertain the validity of an application for family reunification. Rather, it has become an important tipping point in a long, complicated and demanding assessment procedure.[1] This process entails the extensive gathering of narratives and of biometric details about the individuals concerned, as well as interpretation and evaluation of this information according to Danish policies. And, as Feldman has noted in a slightly different context, it involves a complex assemblage of actors, including 'disparate migration policy agendas, generic regulatory mechanisms, and unconnected policy actors and policy "targets"' (Feldman 2011b: 5), the latter, in this case, being refugee families seeking reunification.

In this field of dispersed, disparate and dissociated, yet interlinked actors, I show, no individual person or government agency can be identified as actually taking responsibility for determining whether particular individuals have the right to family reunification. Rather, decision-making has become the responsibility of a 'scientific' system that, in effect, that may end up concealing the actual human decision-making and the policies upon which it is based, thus removing any 'agential blame for traumatic consequences' (Dove 2013: 493). Human decision-making, however, remains central to the procedure and involves a host of small, largely uncoordinated conclusions. Furthermore, it is based on the presumed motives and values that are attributed to the refugees in the Danish political debate and are related to a largely unstated assemblage of ideas about Danish core values and practices that justify the need to control refugees' family lives, even when this may involve questionable, if not illegal practices. The analysis emphasizes that refugees do not live in ordinary human societies subject to the rules and regulations of the nation state to which they belong. Instead, they are left in a kind of extended stateless limbo where the international and national organizations that were ostensibly established to help refugees can be described as having become 'most fundamentally about protecting the global system of national sovereignty by containing and monitoring people out of place because they have fled across an international border [and] not about supporting the rights of refugees to self-determination' (Besteman 2016: 62). In this limbo, refugee families have the choice of either making the best of the marginal existence the large refugee camps can offer them in neighbouring countries – for example, in East Africa or the Middle East – or attempting to flee to more distant countries in Europe or North America, where they may be able to establish a better family life. The latter choice, however, involves struggling with immigration systems that have become increasingly restrictive. And, as already indicated in the introduction to this site, Denmark is a case in point.

Creating and administering a legal apparatus

When the Danish parliament passed a law in 1997 to include biometric verification in the assessment of applications for family relations, this was part of a long process of legal initiatives intended to limit the perceived influx of foreigners and make greater demands on their 'integration' into Danish society. This process is still ongoing, and during the period from 1992 to 2016, no less than 145 changes in legislation were passed by the Danish parliament (Hvidtfeldt and Schultz-Nielsen 2017: 63), entailing, with few exceptions, increasing requirements with regard to immigration, asylum and family reunification. Abdul's flight and family reunification exemplify some of the consequences of these laws for individuals and their families.[2]

Abdul took flight after the militant fundamentalist Muslim group al-Shebaab had arrested him twice because he refused to be recruited for its operations in Somalia. He arrived in Denmark in 2014 with the help of an 'agent' who organized his onward flight and procured the fake passport he needed to enter Europe. When he arrived in Denmark he was fingerprinted, registered as an asylum-seeker and sent to a centre where he would stay while his case was processed. Since he had no formal documentation of his situation in Somalia, he could only substantiate his need for asylum by oral testimony. This took the form of three extensive interviews with officials from the Immigration Service, the first interview involving eight hours of detailed questioning about his identity, place of origin and experiences of persecution. The second interview, conducted more than half a year later, pursued the same kind of questioning but focused on increasingly minute particulars: 'On what day of the week were you arrested [by Al-Shebaab]? What clothes were you wearing? What colour was your shirt?' Abdul found it difficult to remember such facts, which to him were irrelevant to his traumatic experiences and his need for protection. His way of speaking Somali was also recorded so that it could be analysed to determine whether it matched his claimed clan affiliation and place of origin in an area that had been infiltrated by al-Shebaab.

At the third interview, many months later, he was informed that the analysis confirmed his stated subclan and area of origin but did not match the name of the town in Somalia where, according to the first interview, he was supposed to have lived and experienced the events leading to his flight. It turned out the interpreter had mistranslated Abdul's explanation that he came from 'a small town' for the name of a particular city, the two words being similar. Having corrected the translator's mistake, the Immigration Service finally assessed that Abdul's narrative was an accurate, coherent and therefore credible account of persecution (cf. the assessment of plausible stories, Chapter 3), and 15 months after he had submitted his application he was granted temporary asylum. His case would be re-evaluated after five years, and he would be repatriated if conditions in his part of Somalia improved.

When Abdul fled, he lost contact with his family and had no idea for more than a year whether they were dead or alive. But shortly before he was granted asylum, he learned through Somali contacts that his wife and children had fled with his father to a safe area in a neighbouring country. Abdul therefore wanted to apply for family reunification as soon as he received his residence permit. This entailed filling out a 30-page online form for each individual in the family. The local municipality that was responsible for integrating new refugees offered no assistance, but with the help of one of the many volunteer-based services organized by Danish citizens to provide aid to refugees in Denmark, he managed to submit his application just before a new law was passed requiring refugees to wait for three years before being allowed to apply for family reunification.

Almost a year after submitting his application for family reunification, Abdul was informed by the Immigration Service that his wife and children had to be DNA tested at a Danish embassy in order to validate relations of kinship in the family. His family therefore paid an 'agent' to help them undertake an illegal cross-border journey to an embassy in a different country. But just before the testing was to take place, Abdul received a message from the Immigration Service that his asylum was subject to early review because a Danish report had indicated that it was possible to repatriate Somali refugees to certain areas of Somalia that had been declared safe. After another year and a fourth interview, where Abdul was subjected to many of the same questions about the details of his flight, he was re-granted the three remaining years of his asylum status. His wife and children were finally DNA tested and, when the analysis confirmed biogenetic kinship between the children and their parents, the family was reunited some months later. However, Abdul's father, who had fled with Abdul's wife and children and remained with them during the close to four years of the family's separation, had to stay behind alone in the foreign country where the DNA tests were performed, not being considered part of Abdul's nuclear family. 'He can manage now', Abdul said, but 'when he gets older it will be really difficult'.

An extraordinary world

Abdul's story shows how an unholy mixture of the creative negotiation of documents, a convincing presentation of a story of persecution and the risky undertaking of illegal travel has become key to obtaining refuge and family reunification in Denmark. Abdul was not able to travel to Europe without the assistance of an 'agent' (or a 'trafficker') who knew the best route and was able to organize a passport. He had to tell, and retell, a coherent narrative of persecution and flight constructed on the basis of the chaos and trauma he had experienced and maintain the exact same memory of meaningless, incidental details of his flight.[3] Moreover, family reunification required completion of an extensive online application form in a foreign language, another risky and illegal journey by his family organized by yet another 'agent' and the validation through biogenetic testing of his close family relations. In short, in the refugee system the seemingly ordinary depended on the extraordinary: access to legal rights required illegal action, life-threatening experiences were validated by minute, irrelevant details and close, personal family relations were identified and classified biometrically on the basis of swabbed bodily matter.

The routinization of extraordinary situations and conditions within the refugee system was apparent in the interviews I carried out with those who had been DNA tested at the Danish embassy in East Africa. Their family life was characterized by protracted separation caused, to a considerable degree, by the difficulty of fleeing to a country where a normal life might be established and by the extensive bureaucratic procedures set in motion to ascertain the validity of the need for asylum and the right to a family life. The families related harrowing stories of the illegal cross-border travel they had had to undertake without the requisite papers in order to meet the requirement that the DNA testing be performed at a Danish embassy.[4] One family was abandoned by the person who was supposed to help them across the border and had to make their own way in an unknown territory, another family had to go into hiding when the border patrols were getting close and a third family was caught by the patrols, resulting in the mother being arrested. These families were not questioning the DNA test – indeed, they had only vague ideas about what it involved but realized that it somehow had to do with proving their family relations. They were relieved that it was easy and did not hurt and hoped that it signalled the successful completion of a long and difficult process.

A successful DNA test, however, was not the end of testing for all the families. One family, for example, was told to report to the embassy during my research because the Danish immigration authorities suspected that the oldest boy in the sibling group of six might be older than 18 years of age. I followed the boy to a local hospital, where a paediatrician, a radiologist and a dentist made individual age assessments which would be sent to the head doctor, who, I was told, would calculate an estimated age based on international procedures. Age assessments have been criticized for operating with rather wide age ranges and for having a high probability of error (Schmeling et al. 2011; Rinaldi 2016; Abbott 2018; Netz forthcoming). This uncertainty was apparent in an earlier age assessment report, carried out by the same local hospital in connection with another application for family reunification. This showed that the three doctors' individual assessments resulted in an age range of 16 to 23 years, which somehow enabled the head doctor to calculate a final age of 17 to 18 years. Since the Danish immigration authorities have a policy of choosing the lower assessed age, the assessed body was presumably regarded as that of a child who could thus be included in the family's reunification.

As I waited with the boy at different departments in the hospital, it became clear that he had no idea why he had to go through the several examinations and tests. He grew increasingly worried when the various doctors enquired about his general health – to obtain background information on his basic physical condition, I assume – and said to me, 'I am not sick, I am not a problem!' When I asked him how old he was, he replied with confidence that he was 14 years old and that his family had fled when he was eight and a half because there was fighting in the area. During the last two years, he had lived with his aunt and uncle in a neighbouring country to avoid the military draft. The idea that physical tests could establish his age, which would then determine whether he was entitled to be included in a family unit that qualified for reunification in Denmark, was clearly far from his own experiences of family life.

Scientists at the Department of Forensic Medicine of the University of Copenhagen carry out all analyses of DNA tests and perform the biometric age assessments of asylum-seekers in Denmark. During interviews they noted that most of the individuals they tested probably had little knowledge or understanding of the biological science on which these assessments are based and the legal implications of the resultant biogenetic conclusions. In fact, the scientists themselves admitted that they found it difficult to come to terms with the way in which the approximate biological age that they were able to estimate could result in a legal decision:

> It is completely odd to convert a biological age [into a whole number], because it is an ongoing process. You don't grow in notches, but from a legal point of view you do. It is a second past midnight and then, 'Bam!', you have grown a whole year. But biology is not like that, and therefore we always pick the lowest possible age. Even if they end up with a probable age of 16.9, we make a final estimate in which we round it down to 16.

The scientists also realized that it could be problematic to limit the family to the biometrically defined nuclear unit, since the urge to help relatives might extend beyond this narrow biologically determined entity to people who, in practice, are affectively part of the family. This dilemma became apparent, for example, in a court case involving a person who stood accused of having trafficked a boy to Denmark but who claimed he was the father of the child. The DNA test performed by the Department of Forensic Medicine showed that he was probably the child's uncle. Even though it was not standard procedure at the time, the scientists therefore chose to submit a report that not only rejected the man's claim to fatherhood but also included a note stating that he was most likely an uncle. This had the intended effect:

> The judge sat and chewed a little over that story, and then he said, 'Well, obviously he is not the father of the child, but it is also clear that he is the

uncle of the child'. And then he gave him a reduction in the sentence based on the consideration that it is very human to want the best for a nephew. [. . .] Instead of six months' prison he got a three months' suspended sentence.

In their work the scientists thus sought to mitigate the most categorical legal translations of their biometric analyses. At the same time, they regarded biometric tests as the most objective form of assessment available, therefore making their primary responsibility that of ensuring that the tests were carried out as professionally as possible. The rest was up to the Immigration Service.

Determining attachment or detachment in family relations

While biometric testing has become a cornerstone in assessing cases of family reunification, problem cases brought to the voluntary advisory and legal organizations by refugees show that biometric proof is only a precondition, not a guarantee, of family reunification in Denmark. Indeed, a number of additional rules and regulations involving further requirements and restrictions have been instituted to catch unfounded claims for family reunification. One rule concerns the requirement that the parents in a family must be able to prove that they still have an active relationship and that their marriage has not deteriorated during their long period of separation. If this is the case, it is reasoned, there will no longer be any grounds for reunification of the parents – even though it is likely that the long separation and the presumed marital problems were caused by involuntary flight and, to a certain extent, by the cumbersome processing of applications for both asylum and family reunification. The Danish Immigration Service is therefore beginning to demand proof of continued contact between spouses, for example, in the form of transcripts of communication by telephone, email or social media.

One case concerned a family where the husband had been absent from his family for more than five years due to enforced military service, followed by flight and a prolonged stay in a refugee camp in the region and finally the processing of his asylum case in Denmark. When he applied for family reunification 10 months after he had been given refugee status in Denmark, the Immigration Service required information about what kind of contact he had had with his family during his many years of absence and asked why he had waited so long before submitting the application. In his response, prepared by a voluntary legal aid, it was explained that he had been able to visit his family once a year while in military service; that he had had no contact with the family while he was in the refugee camp, there being no telephone and internet service in the area; and that, since arriving in Denmark, he had been able to communicate with his wife and children once every other month, when they undertook a trip of several hours to a relative, where it was possible for them to communicate by phone. It had taken so long before he applied for reunification because he did not know how to go about it and because it was difficult to obtain the documents required by the Immigration Service. The Immigration Service apparently accepted these explanations because reunification was granted quickly after the family had been positively

DNA tested. While the case had a happy ending, thanks to the help of voluntary legal aid, it illustrates how Denmark's requirement to provide documentation proving continued family relations may be out of tune with the refugees' actual conditions of life.

Another rule restricting family reunification in Denmark states that families must live in the country to which they have the greatest attachment, provided conditions in that country are safe (see Rytter 2011, 2013). An East African refugee married to a woman of a different nationality was therefore informed that his family reunification had to take place in his wife's country of origin, even though he had never lived in that country and would experience difficulties making a living there due to conflicts between his own and his wife's countries of citizenship. The Danish authorities nonetheless acknowledged that the children, according to international conventions, had a right to be reunited with their father in Denmark, but their family life there would not include the mother. Being keen to offer their children the best social and economic opportunities, the family opted to have the children, aged five and three, reunified with their father in Denmark, leaving behind the mother in the country where she was a citizen but had not lived for many years. Several months after the children's departure, she applied for a visitor's visa and was interviewed at a Danish embassy about, among other things, the purpose of her trip. She explained that she wanted to be with the family to celebrate her daughter's sixth birthday. A transcript of the English translation of the interview was sent to Denmark, where the Immigration Services would decide whether the visa could be granted.

Finally, the Danish Immigration Service may refuse family reunification if it is suspected that a couple is closely related because this means that the so-called rule of presumption will come into force (see Liversage and Rytter 2015). Arranged marriages between closely related persons, for example cousins, are common in many societies, and within the context of migration, they have been known to play an important role in facilitating the migration of relatives and strengthening transnational family ties. In Denmark, however, arranged marriages are seen to involve strong elements of coercion; indeed, they are often equated with 'forced marriage' (Liversage and Rytter 2015). The Danish rule of presumption thus states that, if a marriage involves two closely related individuals, it is questionable whether both spouses have entered the marriage voluntarily, and a claim for family reunification will therefore be rejected (Familiesammenføringskontoret [Office of Family Reunification] 2009). In one case, a couple suspected of having such a forced marriage had to undergo DNA testing to prove they were *not* related before they were granted family reunification in Denmark.

Closely connected to Danish notions of 'arranged' or 'forced' marriages are suspicions of marriages of convenience, that is, marriages entered into for the sole purpose of obtaining family reunification and thus an immigration visa to a foreign country. It is difficult, however, to make a clear-cut distinction between completely free and entirely arranged marriages. Many young Somalis, for example, are actively dating transnationally through Facebook and other social media motivated by varying interests related to their particular situation as refugees.

During my stay in East Africa, for example, I met with a young, well-educated Somali refugee to give him some clothes that his wife in Denmark had asked me to bring to him. They had found each other on Facebook and dated for three years before they married, shortly after meeting in person in East Africa. After the marriage had been celebrated, they spent a couple of months together, including a one-week honeymoon at a resort, and the wife was pregnant when she returned to Denmark. Their mutual interest in getting married was undoubtedly related to their particular situation as refugees. She was eager to remarry, having lived as a single mother for several years after a failed marriage,[5] whereas he wanted to move to a country where he would be able to further his education and find a good job so that he could support his family. While it was such pragmatic consider- ations that might have led them to surf the internet for a spouse and eventually to marry, this in no way precluded their being fond of each other and wanting to have a family life together. However, the Danish Immigration Service found such cases of family unification highly suspicious and applicants were subjected to detailed questioning. At one such questioning that I witnessed at a Danish embassy, the husband had to respond to questions concerning how he had met his wife, how they celebrated their wedding and whether there were any identifying physical marks on the wife's body, and he was quizzed about his wife's family relations, including the names of particular relatives. When asked for the name of his wife's ex-husband, he responded by exclaiming, 'Why would I be interested in know- ing anything about him?' While the Danish Immigration Service assumed that one would have some knowledge of a spouse's former marital partner, the person interviewed clearly indicated that he was of absolutely no interest to him.

A bureaucratic maze

The preconceived ideas and assumptions about normal family relations held by Danish politicians and public administrators are rarely seriously challenged by the realities of refugee life in East Africa and elsewhere. All decisions related to family reunification are made by the Danish Immigration Service, where applica- tions are processed by teams rather than individual caseworkers. As one frustrated volunteer at a legal service noted, this means that no one team member gains really in-depth information about particular cases and that no individual person is accountable to the refugees, or the public, for the way the cases are handled. Caseworkers rarely meet with any of the relatives involved in family reunifica- tion cases, and contacting the Immigration Service is difficult. Emails, as I myself discovered, may not be answered for many weeks, and phone calls are often put on hold for more than half an hour until a person is available to answer the call. Those who attempt to meet with immigration officials by showing up at the Immi- gration Service will get no further than the reception, which is manned by staff who do not process cases.

Isaias's many visits to the Immigration Service in order to speed up his claim for family reunification exemplifies the bewilderment, frustrations and misunder- standings that refugees may experience when complicated cases are processed according to Denmark's bureaucratic rules by officials who are not in contact with the people they concern. His case file at the Immigration Service, made available to the legal aid volunteer who was assisting Isaias, shows that he made his first trip to the Immigration Service half a year after he had applied for family reunification with his wife, children and an orphaned child of his brother, who had died in a bombing attack. The receptionist made a note in the case that Isaias was upset and worried about his family, explaining they were in a very difficult situation because they were living in a refugee camp 'with poor conditions and robberies'. The receptionist added, 'I advise that we [at the reception] do not handle cases, but that I will forward the information'. The receptionist, however, did notice that some signatures were missing in the application and helped rectify the problem. The processing of the case was prolonged, as certain documents had to be obtained and translated into English or Danish and as the Immigration Service insisted on receiving birth certificates and a marriage license issued by civil authorities, even though it would have been aware that this would be impossible because only religious authorities issue such documents in Isais's country of origin. The case was therefore delayed even further when the religious documents that Isais was able to procure were rejected by the Immigration Service and DNA testing became required. Not only did the family experience difficulty getting to a Danish embassy to have mouth swabs performed, but Isais also did not seem to understand the letter sent to him with information that he also had to be DNA tested. During the close to two years that it took to process this case, Isais made at least 10 visits to the Immigration Service in Copenhagen, where he met nine different receptionists, who wrote notes such as 'Ref. [the Reference] is asking for a decision because he misses his family'; 'He is made aware that a request for DNA has been sent [on a particular date]. He seemed somewhat confused about this information. He is advised to make contact with the [local] hospital'; 'He has come because he is very worried about his family'; 'He is very stressed by the long waiting time'. 'He misses his family, especially his children, and is very worried about them'.

While it is difficult to contact the staff handling family reunification cases at the Immigration Service in Denmark, it is fairly easy for refugees to make contact with the officials at the consular offices at the Danish embassies, where biometric tests and interviews in connection with family reunification are administered. Indeed, I encountered several instances where refugees or non-governmental organizations in Denmark called Danish embassies abroad directly about problems that they could not straighten out with the Immigration Service in Denmark, such as making arrangements to have DNA testing performed at another embassy than the one assigned by the Immigration Service. Consular officers are not only more accessible; they are also more knowledgeable about the conditions of life in the areas from which the refugees originate, and they may show empathy and try to facilitate matters. However, they cannot – and do not attempt to – bend the rules and regulations but perform their work according to the instructions they receive directly from the Danish Immigration Service and under the close supervision of the Danish authorities. Indeed, the Danish supervisor of the consular office at one of the embassies described it as mainly a postbox, a place that receives and delivers the required papers and test material. It was apparent, however, that the locally recruited staff who administer the actual tests and interviews do much more than distribute mail and that they have considerable social and cultural insights that the Danish Immigration Service could have drawn on if it wanted to make its decisions on a more informed basis.

The Immigration Service's elaborate and demanding assessment procedures and the many rejections that were difficult to understand make the whole process of reunification a gruelling one for many families that are in a difficult situation to begin with. In the words of a mother who was DNA tested at one of the Danish embassies I visited,

> You hear different things round about. You hear, among other things, that if you don't have the right papers, they will send you back and the process can be very cumbersome, and this makes you nervous. And my little boy has been very sick lately, and all this about his illness and at the same time having to think about whether I have all the right things has been rather tough for me.

Faced by a largely unfathomable and unapproachable bureaucratic maze that has the power to define the terms of their family life, the refugees choose different approaches to further their case. Some fight to become reunified with those they regard as an important part of their family life. Yousef, for example, insisted that his application for family reunification had to include his 15-year-old nephew, who had lived with the family since he lost his father at the age of one. After a two-year struggle with the Immigration System, Yousef had to give up because he could not provide legally valid proof that the nephew belonged to the family. Others try to adapt their family to what they think the Immigration System expects and will approve. When Najma was interviewed by the Immigration Service to determine whether she could be granted asylum, she decided to say she had only three children, worried that she would be in an unfavourable position if she admitted she had six children, having heard that the Immigration Service was suspicious of very large families. When she received asylum, she wondered how to include all the children in her application for family reunification.

An international no-man's land

The Danish Immigration Service is by no means unique. It is part of a large international refugee 'apparatus' that emerged after World War II and today includes, apart from the UNHCR (the United Nations High Commissioner for Refugees), a collection of more or less coordinated national and international organizations, agencies and government structures that administer and seek to control the swelling population of refugees and asylum-seekers. Originally established to help people in a temporary situation of refuge, the refugee system has acquired increasing significance as it has become apparent that the conflicts that have led people to flee are more permanent than originally envisaged. The refugees who flee to another country basically enter a border world that has become an extensive no man's land inhabited by individuals with limited legal rights, whether as protracted residents of temporary refugee camps, asylum centers or deportation centers or as recognized refugees struggling to establish an ordinary life with their family in a new country, at least temporarily. In this border world, where the extraordinary has become ordinary routine, the general view among refugees is that any headway can only be made through rather dubious means. Those who are stuck in refugee camps may attempt to exploit the social and economic resources of the local refugee systems as much as possible, for example, by adopting an orphan who will be entitled to extra food rations. Others construct exaggerated narratives of their suffering and vulnerability, hoping that this will enable them to be accepted for resettlement or asylum in another country. Yet others engage agents to help them move up north using fake or borrowed documents (Jansen 2008, 2016; Thomson 2012; Besteman 2016: 93–96).

Besteman notes that in this system, which involves people acting out of desperation, organizations working with refugees are often torn between 'on the one hand, the suspicion of refugees as political and economic agents who strategize and conspire, and, on the other, the humanitarian discourse that defines the granting of refuge as moral responsibility and charity' (Besteman 2016: 97–98). One solution to this situation has been the development of an extensive administrative machine in which, as Marnie Jane Thomson expresses it, 'decisions are made within black boxes of bureaucratic opacity' based on mounds of 'files and folders' produced by 'doubting experts' who demand more and more extensive proof from the 'deceptive refugees' (Thomson 2012: 187, 193). Many refugees, Thomson contends (2012: 198), are 'less perturbed by corruption than they are by the months, often years, of being kept in the dark about their cases', often only to 'receive a standard "lack of credibility" rejection letter' for which 'they are usually unable to receive any further explanation from UNHCR representatives'.

In the parallel refugee world of opposing interests, where everything has to be documented and decisions are made by anonymous, opaque bureaucracies, biometric technologies have come to attain a key role. They are seen to produce forms of identification that are 'almost impossible to forge', being based on 'information extracted from our bodies' (Aas 2006: 145), and they are therefore 'represented as infallible and unchallengeable verifiers of the truth about a person – the ultimate guarantors of identity' (Amoore 2006: 343). However, such 'truth' about individuals, as Katja Aas (2006: 155) notes, may bear little relationship to their self-understanding and the lives they have lived. As I have shown, the stipulations as to how biometric tests must be performed, analysed and interpreted in cases of family reunification seem disconnected from the local political and social context in which the refugees exist. The requirement that DNA testing must be performed at a Danish embassy often necessitates illegal and dangerous border crossing. And the very process of DNA testing itself involves a fragmented procedure of mouth swabbing at a Danish embassy abroad, DNA analysis at a hospital in Denmark, and final determination of the right to reunification by the Danish Immigration Service, actors working in different places with limited contact.

Indeed, the biometric results that are produced are interpreted by the Immigration Service within the context of official Danish state definitions of proper family relations, such as the primacy of blood relations, the nuclear family as a natural unit of family life, childhood and family life with the parents ending at the age of 18 and the unnaturalness of marriages between close relatives.

The technical, clinical way in which the biometric tests are performed nonetheless gives the impression that decisions based on the results of such tests are scientific and neutral. As a result, state employees may end up acting according to a scripted authority authored by a faceless state that makes it difficult to take into account humanitarian considerations concerning people who, ironically, are supposedly protected by human rights conventions – as when a mother and wife is not considered to be part of a refugee family life in Denmark or a father or son is left behind alone in a foreign country. As Colin Hoag notes (2011: 84), '[b]ureaucracy and science can act as objectivity machines, generating vision from nowhere and everywhere'. This combination of bureaucratic and scientific approaches thus draws attention away from the cultural and moral assumptions that led to the tests being performed in the first place and towards the conclusions that are drawn from them. The difficulties that these assumptions created for the refugees as they endeavour to re-establish a family life seem less important to the state than the perceived need to safeguard a supposedly threatened, nationally defined Danish culture and society.

Notes

1 Varying policies are followed in different countries, as shown in Heinemann et al. (2013).
2 This case is examined in more detail in Olwig (forthcoming).
3 The expectation that individuals can construct coherent narratives of traumatic experiences has been criticized by anthropologists and others (Norman 2005; Eastmond 2007; Luker 2015; Olwig forthcoming).
4 Other countries have similar requirements, and there is some collaboration between the embassies of different countries.
5 The divorce rate among Somali refugees is high, not only in Denmark but also in other countries in the Global North (see Jagd 2007: 82–88).

Epilogue

In an article published in a BBC travel series on 'Places That Changed the World', Madhvi Ramani (2018) describes the importance of Paris as the place where the metric system was invented in the late 1700s. Drawing on Ken Alder's history of the metric system, *The Measure of All Things* (Alder 2001), she takes the reader to some of the sites in Paris where the efforts to 'metrify' France may be experienced. At the Paris Observatory, a brass strip marks the Paris meridian line used to calculate the length of the metre which was defined as 'one 10-millionth of the distance from the North Pole to the equator'. Below a ground-floor window at the Ministry of Justice, a 'standard metre bar' is still displayed for the public. And the International Bureau of Weights and Measures is home to the 'master platinum standard metre bar' and the original kilogram, the latter placed below 'three bell jars in an underground vault' only accessible 'using three different keys, held by three different individuals'.

Ramani describes the metric system as a great achievement because it established a standardized, universal unit of measurement that could replace the approximately '250,000 different units of weights and measures' then used in France. The French were nevertheless not particularly happy about the metre but wanted to maintain their 'old ways of measuring' that were 'inextricably bound with local rituals, customs and economies', and it therefore took about 100 years before they began to use the metric system. Today, it has been adopted as the standard system of measurement throughout the world, with the exception of the US, Liberia and Myanmar. The metric standardization, Ramani concludes, not only 'formed the basis of our modern economy and led to globalization'; it also 'enabled high-precision engineering and continues to be essential for science and research, progressing our understanding of the universe'.

The metric system not only facilitated the development of biometric technology; it also laid the basis for the idea that various aspects of human life can be measured using highly abstract forms of calculation that may bear no direct relation to this life as experienced by the people that live it. The biometrics of heartbeats, vein structures, fingerprints, DNA analyses and X-rays of teeth and bones that are used to identify, register and categorize individuals bear only a limited relationship to the ways in which people interact, perceive and identify themselves and others in the course of social life. This becomes increasingly apparent as one moves from the identification of travellers through the use of digital facial recognition systems at, for example, the ABC gates at Schengen borders in airports to the registration and categorization of irregular migrants by 'taking their fingerprints' at hotspots in European ports of entry to the determination of family relations by performing DNA analysis on the basis of buccal swabs. As the biometric forms of measurement come to be based on logics that produce a biogenetic mapping of human relations that is removed from everyday life experiences, those who are subject to them have little possibility of understanding and relating to them, even when they concern the most intimate and emotional inter-human relationships – the family. The metrification of society has therefore not only created a uniform standard of measurement that paved the way for scientific and technological discoveries and made possible human interaction on a supra-local and even global level; it has also created a parallel sphere of society that is removed from most people's everyday frames of reference. This development can be related to the shifting political goals behind the introduction of the metric system.

In *The Measure of All Things*, Alder describes how the metre was originally introduced during the French Revolution to create a universal standard of measurement that 'would belong equally to all the people of the world, just as the earth belonged equally to them all' (2001: 1). By abolishing local weights and measures embedded in different social and economic systems that were subject to ongoing negotiation in moral communities of interpersonal relations and replacing them with universal, 'abstracted, commensurable units that relate to an absolute standard' guaranteed by 'a national or international agency' (2001: 126–127), it was believed that it would be possible to create 'an autonomous and egalitarian citizenry able to calculate its own best interests'. However, the political regimes that followed the revolutionary era, such as that of Napoleon, had no time for such 'revolutionary fantasies'. Rather, the metric system was regarded as attractive because it comprised a system of standardization that facilitated the establishment of more centralized rule. The original intended meaning of the system had become irrelevant (2001: 317).

The introduction of national and international biometric systems for purposes of border control can be seen to be part of this movement of political centralization. While it ostensibly employs universal biometric standards that allow individuals to assert their legal rights to asylum within a supranational human rights system, it has been harnessed for narrower national purposes in order to control family migration. As shown in this discussion of the use of biometric technology in assessing cases of family reunification in Denmark, the biometrically verified nuclear family is no longer simply treated as an objective biogenetic unit of people. Instead, it has become a way of protecting a family form that is perceived as basic to Danish national identity and the survival of the Danish nation state. Ironically, it seems that biometrics, like the ancient modes of pre-metric measurements, has become 'inextricably bound with local rituals, customs and economies'. Indeed, as shown by the other ethnographic cases discussed in this monograph, specific biometric technologies, whether conceived and developed in laboratories or put to use at the ABC gates in airports, hotspots at different

Figure IV.2 Seal of the International Bureau of Weights and Measures (Bureau International des Poids et Mesures) representing an allegory of Science holding the new metre standard with its decimal division. On one side of Science is Mercury with geodetic implements (map, globe, compass); on the other an allegory holding a distaff with implements of technology (hammer, anvil, gear). The Greek motto μέτρῳ χρῶ means 'make use of the measure' (or 'make use of the *metre*')

Source: Public domain.

ports of entry or administrative immigration systems, are subject to local human interpretation and application. However, the very idea of its universality as a standardized mode of measurement gives it authority and legitimacy as evidence of the ability of states to protect their borders – and thereby their nationals – in an apparently rational, equitable and just way. But one cannot help but wonder just what a quantitative measure, be it biometric or a measure of the globe that is 'one 10-millionth of the distance from the North Pole to the equator', has to do with human justice?

Conclusion

Karen Fog Olwig, Kristina Grünenberg, Perle Møhl and Anja Simonsen

> *The frontier, an imaginary line connecting mile-stones or stakes, is visible – in an exaggerated fashion – only on maps. But not so long ago the passage from one country to another, from one province to another within each country, and, still earlier, even from one manorial domain to another was accompanied by various formalities. These were largely political, legal, and economic, but some were of a magico-religious nature.*
>
> – van Gennep (1960[1908]: 15)

Rites of passage: the ubiquity of the border

In his seminal work on rites of passage, published in 1908, Arnold van Gennep took his point of departure in rituals connected with passing between different territorial units. In earlier times, he explained, political landscapes were characterized by countries and a host of other more local political units that were not sharply defined. Rather, they were bounded by being 'surrounded by a strip of neutral ground', often thinly populated deserts, marshes or forests, that were attributed with magico-religious significance. Crossing from one territorial unit to another therefore entailed passing through a 'symbolic and spatial area of transition' where the individual 'wavers between two worlds' (1960[1908]: 18–19). However, by undergoing specific rites, or procedures, of passage, it was possible for the individual to complete the transition and thus leave one world and enter another (van Gennep 1960[1908]: 19).

Van Gennep was discussing the phenomenon of territorial passage in relation to a pre-modern political landscape that was quite different from that of the Europe of his time. By the first decade of the twentieth century, when he wrote *Les Rites de Passage*, the frontier formerly dividing countries had become 'an imaginary line connecting milestones or stakes' that was visible 'only on maps' (van Gennep 1960[1908]: 15). Regulations of passage between countries, van Gennep confidently asserted, had become a thing of the past in the civilized world: with the exception of 'a few countries where a passport is still in use, a person in these days may pass freely from one civilized region to another' (van Gennep 1960[1908]: 15).

This statement hardly reflects common sense or a general experience today. In fact, unhindered border crossing seems to have characterized Europe only during the decades from the late nineteenth to the early twentieth centuries. Indeed, this period has been described as 'an unexampled era of free movement in the modern age' (Torpey 2018: 14), when mobility was no longer controlled by local political authorities, such as manorial systems, but had not yet become subject to the control of states (Torpey 2018: 10). With the outbreak of World War I, however, border control became established between European states. While border posts were removed in much of Europe during the 1990s with the passing of the Schengen treaty, travellers had to be in possession of legally valid documents of identification proving their right to be in Europe. Furthermore, as mentioned in the Introduction, border control was tightened on the external borders of the EU, increasingly restrictive policies were introduced to regulate immigrants within Europe, and in 2015 border control was reintroduced between several European countries subscribing to the Schengen treaty. The ability of states to control their borders had become an important aspect of statecraft, borders being regarded as 'the physical manifestation of the sovereignty of the nation and the power of the national state to secure that nation from harm' (Donnan and Wilson 2010: 2). As Torpey notes (2018: 9), this 'monopolization of the legitimate "means of movement"' by states and supra-state systems has been accompanied by the establishment of elaborate bureaucracies that are able to establish individuals' identities and thereby categorize them according to their national identities and their right to enter and live in a particular political territory. Such bureaucracies, he states, depended on what the French historian Gérard Noiriel 'has called the "révolution identificatoire", the development of "cards" and "codes" that identified people (more or less) unambiguously and distinguished among them for administrative purposes' (2018: 9). As the old territorial frontiers between various political entities, including countries, have shrunk to the extent that they have become merely thin lines on the map, they thus have been replaced by more and more extensive administrative systems of centralized control designed to protect the nation state from outside dangers by using increasingly sophisticated biometric technologies.

The ethnographic sites presented in this monograph bear testimony to this increasingly central role played by biometric technologies in border control. Large investments have been made to develop new and more advanced technologies of surveillance and border control, while in biometric laboratories, new forms of biometric identification for use in border control are continuously being explored. These technologies, however, have had very different consequences for different individuals' mobility. They have been developed with the aim of facilitating relatively unhindered, or seamless, border crossing for those whose travel is authorized by documents recognized by the states involved. For most European nationals, entering and leaving the EU is a simple matter of looking sufficiently like the image stored in their official, biometric passports when they are scanned as they pass through the automated border control in the airport. And permission to travel to a country outside Europe will usually not involve more effort than applying and paying for a visa and agreeing to being fingerprinted, or in other ways biometrically registered, upon arrival in the foreign country.

In contrast to the relative ease of travel enjoyed by Europeans, the Africans, Asians and Middle Easterners who attempt to enter Europe to escape from war, conflict and poverty or who just wish to travel on business, attend a conference or go on a holiday will experience border crossing as a daunting task. If they apply for visas to Europe, they must undergo often expensive biometric registration abroad as part of the application. Many will not be able to procure legally valid visas to enter Europe and may therefore attempt to enter without papers. If they reach the land borders in North Africa, they will be closely monitored by cameras so that border guards can prevent attempts to cross the formidable fences. If they try to hide in trucks driving across the border, they may be discovered by heartbeat detectors. If they manage to enter Europe, they will be fingerprinted and registered if they are discovered, whether at hotspots in Southern Europe or at refugee reception centres in Northern Europe, where they may be placed in an asylum system, sometimes for several years, before their case is processed. And if they succeed in acquiring refugee status and wish to become reunited with the families they have left behind, their family members may be subject to DNA testing and possibly age assessments that will delay, and possibly limit, their ability to reunite. The rites of passage that, according to van Gennep, existed to help individuals cross the dangerous borderlands between territorial entities have become ubiquitous technologies of control. Rather than enabling all travellers to leave one world and enter a new one, biometric technologies identify, register, categorize and place some individuals in an in-between world where the legal rights extended to ordinary citizens do not apply.

Routinization of the extraordinary

A prominent characteristic of the biometric border world concerns what we conceptualize as the 'routinization of the extraordinary'. By this we mean that otherwise extraordinary events and experiences become naturalized as ordinary, no longer being notable but becoming part of the daily routine. This takes place, we argue, as the body ceases to be a private matter and is transformed into a publicly accessible object of automated border control and scrutiny. Defined through fingerprints, veins, a beating heart, bones and DNA, the body has been instrumentalized as a locus of identity and institutionalized as an integral part of the biometric border world that can facilitate or constrain movement. Such routinization of the extraordinary occurs, we argue further, when exceptional events, both in the here and now and as imagined future threat scenarios, become integral to the ordinary, everyday work of border control and technological research. This routinization, we show, is institutionalized, expressed and experienced in different ways by the various actors involved.

It is, for example, tacitly accepted that it may only be possible to obtain legal rights to biometrically defined family reunification by illegal means. This is because this process often requires family members abroad – for example, located in camps on the African continent – to embark on illegal border crossing at their own risk, often with the help of brokers (or 'agents'), in order to reach an embassy where DNA tests can be performed. Furthermore, they must accept that foster children who have grown up in the family but whose status within it cannot be proved with legally valid documents or DNA tests have no right to reunification. Recognized refugees and family members claimed by refugees located in Europe have therefore become trapped in huge camps in Africa and the Middle East where they live in a legally accepted state of limbo for years on end.

Routinization of the extraordinary is also apparent in the treatment of migrants' bodies as suspicious moving objects that the border police have the right to inspect and enrol or register at any given time or place. As explained by a rejected asylum-seeker, his body had been turned into a public border zone that no longer had any boundaries of privacy. The hotspots in Italy are examples of border zones where the body of people *en route* is by definition regarded as suspect and therefore turned into a public site of inspection.

For the guards conducting border inspections, the search for the extraordinary has become part of the routine of their work. Using a combination of biometric technologies, tall fences monitored by video cameras, heat and sound detectors, and their own vision and sixth sense, the detection of what might in other contexts be experienced as extraordinary – a heartbeat in the bottom of a truck, what looks like a pistol in a suitcase, even hundreds of migrants storming a border fence in joined force – has become an integral aspect of routine border work. By virtue of the individual discretion they are endowed with, border guards are left with the responsibility for maintaining an acceptable balance between open borders and total closure, as well as between profiling based on appearances and the legally required ethical norms of non-discrimination.

Likewise, when researchers and developers of technology try to imagine potential new ways of circumventing biometric technologies, such as by using fake fingers, or when they experiment with parts of the body or physiological processes such as heartbeats that have the potential to be used for presence detection and biometric identification and recognition, imaginative extraordinary scenarios for the future become part of a normal laboratory work routine. Potential scenarios, in other words, are tested scientifically in order to determine whether and how new technologies may be employed for purposes of identification and authentication. The extraordinary also becomes a vital driving force in laboratory researchers' daily undertakings when they participate in border conferences and funding negotiations where the imagining of new threats takes place, as exemplified by the regular evocation of that *one* incident of a person trying to pass the border wearing a rubber mask.

As these examples demonstrate, the biometrification of border control has resulted in the development of an extensive, technologically sophisticated and costly border world administered by vast national, supranational and international bureaucracies that are materialized in the practices and decisions of particular border guards, migrants and researchers. Such practices and decisions are often contradictory and not systematically connected to explicit and coherent administrative, political and economic agendas. Despite the high price – social, emotional and financial – European nation states regard biometric border control as of the utmost importance as a scientifically based, reliable and coordinated system that can protect European countries from external threats.

Biometric border control, as we have demonstrated, is indeed based on careful, painstaking, imaginative research conducted by scientists and engineers, and it can be used to recognize individuals who are fingerprinted or otherwise biometrically registered in databases, to detect human presences and movements in important areas of border control and to establish biological relationships of kinship. But it is not foolproof in terms of its ability to identify and recognize. Furthermore, and most significant of all, it depends heavily on human interpretation, knowledge and sensitivity to particular situations and conditions of mobility and flight. However, the human factor is often obscured by the supposed scientific validity of biometric technology as a guarantor of truth. In this manner, human responsibility for the ways in which policies and administrative practices are implemented has therefore become hidden from view and turned into a routinization of the extraordinary.

Assemblages, connections and disconnections

It is not possible to study 'the border world' in and of itself. As Bateson reminds us, '[w]hat can be studied is always a relationship or an infinite regress of relationships, never a thing' (Bateson 1989: 246). Throughout the book, therefore, rather than a ready-made object of enquiry, we have operated with the idea of the 'border world' as an assemblage constituted by ephemeral and contingent relations between a variety of different places, actors, socio-material practices and small-scale everyday decisions made in our respective sites, as well as the particular ways in which the economy, politics, affects and desires resonate in those decisions. In our writing, then, 'the border world' figures as both an analytical strategy and a definition of an empirical field in the making.

The small decisions made by the different actors in our sites at the level of human interaction resemble 'tactics' as described by Michel de Certeau (1984) in that they take the form of 'making do'. De Certeau defines 'making do' as the adjustments made to particular environments and situations as they emerge but without the presence of an overall strategy operating from what he calls a 'proper', that is, an external, visible and distinct 'spatial or institutionalized localization' (de Certeau 1984: xix). We have shown that, although the manifold decisions and varied elements and actors involved in border work at times become connected and produce very tangible effects, both in and across sites, in practice they do not operate strictly according to an underlying organizing principle, nor are they expressions of any sort of coherent master plan or overall strategy. If anything, there are many, more or less coherent and sometimes colliding, strategies or plans. The implementation of the EU smart (biometric) border system, the Dublin agreement, the biometric EURODAC database and the adoption of DNA testing to define 'the family' in many European countries constitute extensions and continuations of an increased focus on security and migration. Such systems are indeed intended to turn borders into increasingly fine-grained filtering mechanisms, and they have important implications for those who are attempting to migrate or to become reunited with their families, as well as for the work taking place at borders and in laboratories. As this book has shown, however, these systems do not simply predetermine border work on the ground. Instead, the decision-making work that takes place in the four sites we have described and analysed in the foregoing chapters is often the outcome of uncoordinated and unplanned tactics, of local and/or contextual circumstances and coincidences.

Such decision work includes occasions when migrants make use of 'in-formation' from Somali sources in order to avoid biometric registration and continue their journey through Italy, when a vague white figure is defined by border guards in Ceuta as a donkey and not a human or as a Maghrebi and not a sub-Saharan individual, when the use of DNA tests to establish the right to family reunification ends up being appropriated by some Somalis to negotiate their transnational family obligations and when researchers manage to procure funding for research that will lead to the development of a particular new kind of biometric technology because they choose to attend one biometric event rather than another.

Apart from such instances of contextual tactics, circumstances and small decisions, our border-world assemblage is composed of ephemeral and mutable connections. The types of biometric technologies developed in the labs, for example, are presently connected to a threat scenario that defines danger as coming from the outside. This type of connection is commercially profitable and can generate significant funding for research and development of new technology. The connection, however, may change when the threat landscape mutates (due, for example, to climate change or disease epidemics) or when decisions to change an algorithm leads to new results, making it possible to use biometrics to address other types of 'threat', such as those posed by illness. Such developments may facilitate access to other sources of funding, thereby diverting interest away from borders. Along the same lines, individuals' decisions about migration trajectories, which are often made by connecting bits of information from a variety of shifting sources, will be altered as migrants acquire and put together new bits of information about changes in the border world, such as the establishment of a new 'hotspot' and the installation of new types of biometric gear at a particular border crossing. Just as such reconfigurations mean that the connections or relations that partially constitute the border world assemblage are subject to ongoing readjustment, so they may also bring about *disconnections*. The practices of border guards, for example, rather than being the 'extended arm of the state', are often partly disconnected from governmental and EU directives, as well as from the commercial interests that also regulate the conditions of border control. This becomes evident when, for example, Frontex directives require border guards to identify a certain number of so-called high-risk passengers against the guards' perceptions of right and wrong, as well as EU laws prohibiting racial profiling. In addition, biometric gates turn into obsolete pieces of machinery disconnected from the border world when they fail to work or are simply switched off by border guards at the airports because the installations 'get confused' by the light or by travellers with lots of hand luggage. Disconnections are also at play in the labs when body topologies or patterns, like vein patterns, are disconnected from the living bodies from which they were originally harvested and reconnected to vein patterns from other bodies in the production of usable data sets. Another type of disconnection is at play when migrants are fingerprinted at the Italian border and thereby legally disconnected from their desired trajectories or when the Danish family reunification system purportedly reconnects families torn apart by wars and catastrophes, whereas DNA and bone scans in practice maintain the disconnections between some family members in the name of an ideal nuclear family model.

In sum, the notion of assemblage highlights the changeable, fleeting, incoherent and moving parts of the border world, as well as the social, political and economic forces that shape and are shaped by it. Where connections at times seem to link the parts as well as the different actors in the biometric border world, whether they be researchers, border guards, politicians or migrants, just as often voids and disconnections become perceptible, leading the analysis towards phenomena and ways of thinking and acting of a more intangible nature.

Wishful thinking and faith in the 'technological sublime'

The many procedural, technological and material decision-making formalities operating in, and conditioning, the border world are generally, as van Gennep proposes, defined in 'largely political, legal, and economic' terms. This is essentially how they are described in national and European policy papers, by the media, in critical research and by politicians and bureaucrats. The fingerprint, the family reunification form, the visa, the camera, DNA analysis, the scanner, the vein pattern and the swab are therefore, to many people, the necessary formalities of border work, anchored in common sense, even if to others they can appear opaque, incoherent and disconnected. In the context of migration and mobility regulation, they thus ostensibly provide concrete solutions to concrete threats against national and European interests and security, including threats seen to derive from migration.

Yet throughout our fieldwork and writing, we have sensed the presence of something else, of another order at play in the border world – something more intangible and evanescent that we have at times described as imaginary, mythical, potential or as wishful thinking, depending on the perspective. We argue that these more intangible phenomena and forces play just as important a role in the processes characterizing the border world and the small decisions that give it substantiality as do the purportedly tangible political, economic, legal and material impetuses. Although, as van Gennep suggests, we can characterize such intangible imaginary impetuses as 'formalities . . . of a magico-religious nature', our analyses nevertheless show that they do not exist outside and apart from political, legal and economic formalities. On the contrary, they are just as integral to political, legal and economic decision-making as they are to making mundane daily choices. These types of imaginary and wishful thinking are not of a lower explanatory order, relegated to primitive thinking as opposed to scientific forms of logic (cf. Jöhncke and Steffen 2015: 19). Indeed, in the assemblages made up of the many apparently rational administrative procedures, technological activities, scientific procedures and technologically assisted or performed decision-making procedures that we have described and analysed throughout the foregoing chapters, the imaginary and intangible play an important role in establishing, maintaining and also circumventing borders.

The enmeshing of the imaginary and the intangible within rational political, legal and economic 'formalities' has been pointed out in several studies. In *On the Modern Cult of the Factish Gods*, Bruno Latour describes how colonial Portuguese envoys encountering a Guinean people and their unwavering faith in the powers of their divine figurines ask them how they can possibly believe in the divine powers of something they have themselves made (2010: 3). Latour poses the same question about Western scientific facts and rationality. Pointing to the example of Louis Pasteur, who pictures himself producing facts in his laboratory, Latour asks whether these facts are constructions or real, cultural or natural (2010: 16)? Michael Herzfeld (1993: 17) makes a similar move when he criticizes the widespread presumption that 'the bureaucratically regulated state societies of "the West" are more rational – or less "symbolic" – than those of the rest of the world'. This idea, he states, 'treats rationality as distinct from belief, yet demands an unquestioning faith not radically different from that exacted by some religions'. The notion of the perfectly ordered state and bureaucratic reason therefore has 'the marks of a religious doctrine'; reason is a symbolic construct, whether applied to state bureaucracy or to scientific procedures and explanations (Herzfeld 1993: 17). According to these approaches, science and bureaucracy, technologies and state apparatuses are governed by an interlacing of rational and religious formalities, as also proposed by van Gennep. But how does that relate to the imaginaries and the forms of wishful thinking described earlier? And how do the magico-religious formalities and aspects of biometric technologies relate to the practical instantiations of the biometric border world that we have encountered in the field?

We suggest that this confidence in technological solutions is related to what David Nye has defined as 'the technological sublime', that is, the feeling of awe that modern technologies have generated as extraordinary human inventions believed to be manifestations of human reason (Nye 1994: 60). There is an element of faith, as well as wishful thinking, involved when policy-makers expect systems operating with vein patterns to be objective, to work flawlessly and to produce 100% foolproof results, in spite of laboratory researchers' knowledge of potential database biases, errors, systemic glitches and the potential for fooling such systems. At the same time, these researchers are motivated by a continually sustained aspiration to find that foolproof technology, that inimitable body part – an ambition that nourishes every daily decision, down to the writing of the smallest piece of code.

There is unshakable confidence and a desire for control at play when it is assumed that there exists a direct and enduring connection between a small ID photo of a face and a human life in all its complexity, a connection that, when confirmed through masterful mimicry, opens the border. The idea that 'we know who you are' is based on very shaky grounds and, one could say, on wishful thinking and a professed faith in, even fascination with, the superhuman 'digital sublime' (Mosco 2004). There is a dream of total, objective control when the biometric ABC gate takes over the border work and the role of selective filtering, unhampered, as it is imagined to be, by human biases, fatigue and resistance to managerial decisions. It is above and beyond that man-made Guinean fetish, a supra-human agent that is disconnected from human intervention and responsibility. And the dream of the impenetrable border itself, in its many 'smarter', 'intelligent' and highly technologized forms, is in itself a kind of wishful thinking invoked both to obstruct the proposed threats of unending waves of migration and to further political agendas. The suggestion that images recorded and transmitted by drones may be the solution to border control bears an unmistakable tinge of faith in the digital/technological sublime, including in the eyes of the border guards who do not hold the belief.

There is a digitally founded faith in the fingerprint as the beholder of the truth about a person's identity. To the person carrying that fingerprint, finding ways to resist is often either fruitless or self-destructive. Yet at times, ways of circumventing the EURODAC database, and thereby the Dublin regulations, appear when by happenstance one's fingerprint is *not* 'taken' because there is a gap in the fence or in the border control because a police officer cares more for a state than for the EURODAC system. And new avenues forward may emerge when one's information networks have yielded pathways and loopholes that turn out to be efficient. The future itself is, for a migrant *en route*, the beholder of all wishful thinking, of all the imagined fortunes of a secure and peaceful life that can be realized by making the right decisions.

Indeed, it could be characterized as belonging to the order of wishful thinking that a humid swab can lead to the reunification of family members across wide stretches of land and sea. The relationship between cause and effect may appear completely obscure to the family member whose mouth is brushed by a tiny swab that may, by some unknown procedure and logic, procure a much-wanted laissez-passer. But from another perspective, to the operator producing the DNA sample, to the scientist starting the DNA sequencing and harvesting the results, to the administrative agent in an office in Copenhagen who can simply take notice, the test has an uncanny touch of magic in its objective, evidential nature, which seems to be beyond human intervention, decision-making and thus responsibility.

The wishful thinking and faith that are an integral part of the political, legal and economic 'formalities' of border crossing must be seen in light of the disconnection and uncertainty that characterize the border world. On our ethnographic journey through our field sites, we have thus uncovered a strong element of disconnection where the relationship between cause and effect, or intentions and outcomes, has seemed opaque, absurd or questionable. This appears to have generated a heightened sense of uncertainty that has led to continued efforts to find new ways to counter potential threats, whether in the form of new ways of circumventing the border or new technologies that can make the border more effective. In this border world, we suggest, biometric technologies, with their quasi-magical properties, have been attributed powers of both control and foresight that generate fascination and bottomless confidence, as well as bewilderment and estrangement, depending on the perspective from which they are experienced.

We therefore argue that, while the border world is administered and regulated by various national and supranational agreements and conventions that are closely associated with influential political, legal and economic interests, these official formalities merely constitute a fairly tenuous framework that is marked by unrecognized 'formalities . . . of a magico-religious nature' of cross-border action and interaction. Biometric technologies, our study has shown, constitute a central dynamic in the contemporary border world. Not only have they contributed to extending the spatial and temporal sphere of border control. In the process of placing 'unwelcome' border-crossers within in-between worlds located outside the national order and characterized by a routinization of the extraordinary, they have also proffered a mode of border control that is associated with efficiency and objective treatment. These technologies have thereby nourished wishful thinking about the states' ability to maintain a world that is both orderly and fair. In the process, however, this has required the animate and mobile biological being of the human body and its cultural identities to be metricized, technified and digitalized in order to be identifiable within the inanimate realm of the bordered and frozen geometric space of the modern nation state.

References

Aas, Katja Franko. 2006. 'The body does not lie': Identity, risk and trust in technoculture. *Crime, Media, Culture* 2(2): 143–158.

Abbott, Alison. 2018. DNA clock may aid refugee age check. *Springer Nature* 561: 15.

Abrams, Kerry and R. Kent Piacenti. 2014. Immigration's family values. *Virginia Law Review* 100(4): 629–709.

Alder, Ken. 2001. *The Measure of All Things: The Seven-Year Odyssey and Hidden Error that Transformed the World*. New York: The Free Press.

Alpes, Maybritt. 2015. Airport casualties: Non-admission and return risks at times of internalized/externalized border controls. *Social Sciences* 4(3): 742–757.

AMISOM. 2019. http://amisom-au.org/amisom-background/ (accessed February 4, 2019).

Amoore, Louise. 2006. Biometric borders: Governing mobilities in the war on terror. *Political Geography* 25: 336–351.

Amoore, Louise. 2009. Taking people apart: Digitised dissection and the body at the border. *Environment and Planning D: Society and Space* 27: 444–464.

Amoore, Louise. 2013. *The Politics of Possibility: Risk and Security Beyond Probability*. Durham: Duke University Press.

Anand, Nikhil. 2017. *Hydraulic City: Water and the Infrastructures of Citizenship in Mumbai*. Durham: Duke University Press.

Andersen, Marie-Louise. 2018. Somalisk familie der gik under jorden er tilbage i Odder. *Aarhus Stiftstidende*. December 13, 2018. https://jyllands-posten.dk/aarhus/odder/ECE1 1071098/somalisk-familie-der-gik-under-jorden-er-tilbage-i-odder/ (accessed February 20, 2019).

Andersson, Ruben. 2010. Wild man at Europe's gates: The crafting of clandestines in Spain's Cayuco crisis. *Etnofoor* 22: 31–49.

Andersson, Ruben. 2014. *Illegality, Inc.: Clandestine Migration and the Business of Bordering Europe*. Berkeley: University of California Press.

Andersson, Ruben. 2016. Hardwiring the frontier? The politics of security technology in Europe's 'fight against illegal migration'. *Security Dialogue* 47(1): 22–39.

Appadurai, Arjun. 1996. *Modernity at Large: Cultural Dimensions of Globalization*. Minneapolis: University of Minnesota Press.

Armbruster, Heidi and Anna Laerke, eds. 2008. *Taking Sides: Ethics, Politics and Fieldwork in Anthropology*. Oxford: Berghahn Books.

ASGI: The Associazione pergli Studi Giurdici Sull'Immigrazione. 2019. www.asylumi neurope.org/reports/country/italy/asylum-procedure/general/short-overview-asylum-procedure (accessed February 14, 2019).

Augé, Marc. 2008. *Non-Places: Introduction to an Anthropology of Supermodernity*. London: Verso.

Baird, Theodor. 2016. Who speaks for the European border security industry? A network analysis. *European Security* 26(1): 37–58.

Baird, Theodore. 2017. Knowledge of practice: A multi-sited event ethnography of border security fairs in Europe and North America. *Security Dialogue* 48(3): 187–205.

Baldassar, Loretta. 2016. De-demonizing distance in mobile family lives: Co-presence, care circulation and polymedia as vibrant matter. *Global Networks* 16(2): 145–163.

Barata, Llilda P., Helene Starks, Maureen Kelley, Patricia Kuszler and Wylie Burke. 2015. What DNA can and cannot say: Perspectives of immigrant families about the use of genetic testing in immigration. *Stanford Law & Policy Review* 26: 597–638.

Barth, Fredrik. 1969. Introduction, pp. 9–39 in Fredrik Barth, ed. *Ethnic Groups and Boundaries: The Social Organization of Cultural Difference*. Boston: Little, Brown and Company.

Bateson, Gregory. 1989. *Steps to an Ecology of Mind: A Revolutionary Approach to Man's Understanding of Himself*. San Francisco: Chandler Publishing Company.

Bauer, Daniel. 2014. Reflections on roles. *The Applied Anthropologist* 34(1–2): 10–16.

Becker, Heike, Emile Boonzaier and Joy Owen. 2005. Fieldwork in shared spaces: Positionality, power and ethics of citizen anthropologists in Southern Africa. *Anthropology Southern Africa* 28(3–4): 123–132.

Bellacasa, María Puig de la. 2012. 'Nothing comes without its world': Thinking with care. *Sociological Review* 60(2): 197–216.

Benson, Jillian and Jan Williams. 2008. Age determination in refugee children. *Australian Family Physician* 37(10): 821–824.

Berger, John. 1982. The ambiguity of the photograph, pp. 85–100 in John Berger and Jean Mohr, eds. *Another Way of Telling*. New York: Pantheon.

Bernard, H. Russell. 2013. *Social Research Methods: Qualitative and Quantitative Approaches*. Los Angeles: SAGE Publications.

Besteman, Catherine. 2016. *Making Refuge: Somali Bantu Refugees and Lewiston, Maine*. Durham: Duke University Press.

Besteman, Cathrine. 2017. Experimenting in Somalia: The new security empire. *Anthropological Theory* 17(3): 404–420.

Bigo, Didier. 2002. Security and immigration: Toward a critique of the governmentality of unease. *Alternatives* 27(1): 63–92. Sage Publications.

Bindslev, Marianne With. 2004. Færre snyder om familieforhold efter dna-tjek. *Kristeligt Dagblad*. March 4, 2004. www.kristeligt-dagblad.dk/danmark/frre-snyder-om-familie forhold-efter-dna-tjek (accessed March 17, 2017).

Biometric Institute. 2015. *Is a Photo a Biometric?* www.biometricsinstitute.org/faq-16 (accessed March 28, 2015).

Boas, Franz. 1893. Remarks on the theory of anthropometry. *Publications of the American Statistical Association* 3(24): 569–575.

Bogard, William. 2006. Welcome to the society of control: The simulation of surveillance revisited, pp. 55–78 in Kevin D. Haggerty and Richard V. Ericson, eds. *The New Politics of Surveillance and Visibility*. Toronto: University of Toronto Press.

Boulamwini, Joy and Timnit Gebru. 2018. Gender Shades: Intersectional accuracy disparities in commercial gender classification. *Proceedings of Machine Learning Research* 81: 1–15.

Breckenridge, Keith. 2014. *Biometric State: The Global Politics of Identification and Surveillance in South Africa, 1850 to the Present*. Cambridge: Cambridge University Press.

Brekke, Jan-Paul and Grete Brochmann. 2014. Stuck in transit: Secondary migration of asylum seekers in Europe, national differences, and the Dublin Regulation. *Journal of Refugee Studies* 28(2): 145–162.

Broeders, Dennis. 2007. The new digital borders of Europe: EU databases and the surveillance of irregular migrants. *International Sociology* 22(1): 71–92.

Burt, Chris. 2018. *Scientists Develop New 'Brain Password' Biometric Using Electrical Response to Images*. June 7, 2018. Biometrics Research Group, inc. www.biometricupdate. com/201806/scientists-develop-new-brain-password-biometric-using-electrical-response-to-images (accessed April 5, 2019).

Capitani, Giulia. 2016. *Oxfam Briefing Paper: Hotspot, Right Denied*. May 2016. pp. 1–40. www.oxfam.org/sites/www.oxfam.org/files/file_attachments/bp-hotspots-migrants-italy-220616-en.pdf (accessed June 19, 2018).

Carsten, Janet. 2000. Introduction: Cultures of relatedness, pp. 1–36 in Janet Carsten, ed. *Cultures of Relatedness: New Approaches to the Study of Kinship*. Cambridge: Cambridge University Press.

Certeau, Michel de. 1984. *The Practice of Everyday Life*. Berkeley: University of California Press.

Chandler, Daniel. 2014. Icons and indices assert nothing, pp. 131–136 in Torkild Thellefsen and Bent Sorensen, eds. *Charles Sanders Peirce in His Own Words: 100 Years of Semiotics, Communication and Cognition*. Berlin and Boston: De Gruyter.

Citizensinformation.ie.en (Citizens Information Board). 2019. www.citizensinformation. ie/en/moving_country/asylum_seekers_and_refugees/the_asylum_process_in_ireland/ dublin_conventionhtml (accessed February 14, 2019).

Cohn, Simon. 2010. Seeing and drawing: The role of play in medical imaging, pp. 91–105 in Cristina Grasseni, ed. *Skilled Visions: Between Apprenticeship and Standards*. Oxford: Berghahn Books.

Collier, Stephen J. 2006. Global assemblages. *Theory, Culture & Society* 23(2–3): 399–401.

Collyer, Michael, Franck Düvell and Hein de Haas. 2012. Critical approaches to transit migration. *Population, Space and Place* 18: 407–414.

Consuegra, Augustin Diaz De Mera. 2019. Legislative train schedule: Towards a new policy on migration. *The European Parliament*. www.europarl.europa.eu/legislative-train/ theme-towards-a-new-policy-on-migration/file-entryexit-system-(2016-smart-borders-package) (accessed April 4, 2019).

Council of Europe. 2018. *European Convention on Human Rights*. European Court of Human Rights. www.echr.coe.int/Documents/Convention_ENG.pdf (accessed March 16, 2018).

Coutin, Susan Bibler. 2003. *Legalizing Moves: Salvadoran Immigrants' Struggle for U.S. Residency*. Ann Arbor: University of Michigan Press.

Cregan, Kate. 2007. Early modern anatomy and the queen's body natural: The sovereign subject. *Body and Society* 13(2): 47–66.

Cunha, Eugénia, Eric Baccino, Laurent Martrille, Frank Ramsthaler, Joana Prieto, Yves Schuliar, Niels Lynnerup and Cristina Cattaneo. 2009. The problem of aging human remains and living individuals: A review. *Forensic Science International* 193(1–3): 1–13.

Cutler, Victoria and Susan Paddock. 2009. Use of Threat Image Projection (TIP) to enhance security performance. *Proceedings – International Carnahan Conference on Security Technology*: 46–51.

Dalgas, Karina Märcher. 2015. Becoming independent through au pair migration: Self-making and social re-positioning among young Filipinas in Denmark. *Identities: Global Studies in Culture and Power* 22(3): 333–346.

David, Matthew. 2002. Problems of participation: The limits of action research. *International Journal of Social Research Methodology* 5(1): 11–17.

The Day Book. 1912. The Chinese way. *Chronicling America: Historic American Newspapers* (Chicago, IL). July 31, 1912, image 15. Library of Congress. https://chroniclin gamerica.loc.gov/lccn/sn83045487/1912-07-31/ed-1/seq-15/ (accessed April 4, 2019).

De Genova, Nicholas. 2013. Spectacles of migrant 'illegality': The scene of exclusion, the obscene of inclusion. *Ethnic and Racial Studies* 36(7): 1180–1198.

Dekker, Rianne, Godfried Engbersen, Jeanine Klaver and Hanna Vonk. 2018. Smart refugees: How syrian asylum migrants use social media information in migration decision-making. *Social Media + Society*. January–March: 1–11.

De León, Jason. 2015. *The Land of Open Graves: Living and Dying on the Migrant Trail.* Berkeley: University of California Press.

Deleuze, Gilles. 2003. *Francis Bacon: The Logic of Sensation.* London: Continuum.

Deleuze, Gilles and Félix Guattari. 1987. *A Thousand Plateaus: Capitalism and Schizophrenia.* Minneapolis: University of Minnesota Press.

Denková, Adéla. 2016. Frassoni: Italians believe that the EU abandoned them to the migration crisis. *EUROACTIV.* October 2016. www.euractiv.com/section/elections/interview/frassoni-italians-are-convinced-the-eu-abandoned-them-in-migration-crisis/ (accessed February 14, 2019).

Diiriye, Axmed Cali, Bashiir Saleebaan Aadan (B. Raadceeye), Cabdiqani Ibrahin Gaas, Nuuradiim Cabdiraxmaan Sh Abukar, Saamiya Mahad Fadal, Saleebaan Maxamuud Jaamac and Xamse Maxamed Cali. 2015. *Magafe, Tahriibka iyo Dhallinta Sibiq-dhaqaaqday.* Somaliland: Nahda.

Dijstelbloem, Huub and Dennis Broeders. 2015. Border surveillance, mobility management and the shaping of non-publics in Europe. *European Journal of Social Theory* 18(1): 21–38.

Dobbs, Erica and Peggy Levitt. 2017. The missing link? The role of sub-national governance in transnational social protections. *Oxford Development Studies* 45(1): 47–63.

Dodge, Martin and Rob Kitchin. 2004. Flying through Code/Space: The real virtuality of air travel. *Environment and Planning A* 36(2): 195–211.

Donnan, Hastings and Dieter Haller. 2000. Liminal no more. *Ethnologia Europaea* 30(2): 7–22.

Donnan, Hastings, Bjørn Thomassen and Harald Wydra. 2018. The political anthropology of borders and territory: European perspectives, pp. 344–359 in Harald Wydra and Bjørn Thomassen, eds. *Handbook of Political Anthropology: Elgar Handbooks in Political Science.* Cheltenham: Edward Elgar Publishing.

Donnan, Hastings and Thomas M. Wilson. 1999. *Borders: Frontiers of Identity, Nation and State.* Oxford: Berg.

Donnan, Hastings and Thomas M. Wilson. 2010. Ethnography, security and the 'frontier effect' in borderlands, pp. 1–20 in Hastings Donnan and Thomas M. Wilson, eds. *Borderlands: Ethnographic Approaches to Security, Power, and Identity.* Lanham: University Press of America.

Dourish, Paul. 2016. Algorithms and their others: Algorithmic culture in context. *Big Data & Society* 3(2): 1–11.

Dove, Edward S. 2013. Back to blood: The sociopolitics and law of compulsory DNA testing of refugees. *UMass Law Review* 8(2): 466–530.

Düvell, Franck and Bastian Vollmer. 2011/01. Improving US and EU Immigration systems: European Security Challenges. *EU-US Immigration Systems.* The Robert Schuman Centre for Advanced Studies, San Dominico di Fiesole (FI): European University Institute.

http://cadmus.eui.eu/bitstream/handle/1814/16212/EU-US%20Immigration%20Sys tems2011_01.pdf?sequence=1 (accessed March 18, 2019).

Dynetics. 2018. *GroundAware: Protecting Critical Infrastructure and Perimeter Security*. www.dynetics.com/groundaware/ (accessed July 5, 2018).

Eastmond, Marita. 2007. Stories as lived experience: Narratives in forced migration research. *Journal of Refugee Studies* 20(2): 248–264.

Echarri, Carmen. 2017. Entrada a la carrera por la frontera del Tarajal. *El Faro de Ceuta*. https://elfarodeceuta.es/entrada-la-carrera-la-frontera-del-tarajal/ (accessed August 8, 2017).

Ejrnæs, Morten. 2001. Integrationsloven – en case, der illustrerer etniske minoriteters usi-kre medborgerstatus. *AMID Working Paper Series* 1/2001, Aalborg University. http:// vbn.aau.dk/files/33970539/01_ejrnaes.pdf (accessed April 1, 2019).

El Faro de Ceuta. 2017. Grabación de la entrada por el Tarajal de Casi 200 Inmigrantes. *El Faro de Ceuta*. https://elfarodeceuta.es/grabacion-la-entrada-tarajal-casi-200-inmigrantes/ (accessed August 8, 2017).

El Faro de Ceuta. 2018. *El Balance de La Inmigración En 2017*. https://elfarodeceuta.es/ balance-2017-inmigracion-ceti/ (accessed March 2, 2019).

EU Migration and Home Affairs. 2019. https://ec.europa.eu/home-affairs/what-we-do/ policies/borders-and-visas/smart-borders_en (accessed April 3, 2019).

European Commission. 2000. *Charter of Fundamental Rights of the European Union*. Bruxelles: Official Journal of the European Communities.

European Commission. 2014. *Migration and Home Affairs – Smart Borders*. https:// ec.europa.eu/home-affairs/what-we-do/policies/borders-and-visas/smart-borders_en (accessed October 3, 2018).

European Commission. 2015a. *Communication From the Commission to the European Par- liament, the Council, the European Economic and Social Committee and the Commit- tee of the Regions: A European Agenda on Migration*. https://ec.europa.eu/home-affairs/ sites/homeaffairs/files/what-we-do/policies/european-agenda-migration/background- information/docs/communication_on_the_european_agenda_on_migration_en.pdf (accessed April 5, 2019).

European Commission. 2015b. *Asylum in the EU*. September 9, 2015. https://ec.europa.eu/ home-affairs/sites/homeaffairs/files/e-library/docs/infographics/asylum/infographic_ asylum_en.pdf (accessed February 14, 2019).

European Commission. 2016. *Annex 4*. March 16, 2016. https://ec.europa.eu/home-affairs/ sites/homeaffairs/files/what-we-do/policies/european-agenda-migration/proposal- implementation-package/docs/20160316/first_report_on_relocation_and_resettle- ment_-_annex_4_en.pdf (accessed February 14, 2019).

European Commission. 2017a. *Commission Recommendation of 12.5.2017 on Proportion- ate Police Checks and Police Cooperation in the Schengen Area*. http://eur-lex.europa. eu/LexUriServ/LexUriServ.do?uri=COM:2012:0230:FIN:EN:PDF (accessed June 16, 2017).

European Commission. 2017b. *Family Reunification of Third-Country Nationals in the EU Plus Norway: National Practices*. EMN Synthesis Report for the EMN Focussed Study 2016. Migrapol EMN [Doc 382]. https://ec.europa.eu/home-affairs/sites/homeaffairs/ files/00_family_reunification_synthesis_report_final_en_print_ready_0.pdf (accessed January 16, 2019).

European Commission. 2018. *Managing Migration: Commission Expands on Disembar- kation and Controlled Centre Concepts*. July 24, 2018. http://europa.eu/rapid/press- release_IP-18-4629_en.htm (accessed February 12, 2019).

European Council. 2016. *EU-Turkey Statement*. March 18, 2016. www.consilium.europa.eu/en/press/press-releases/2016/03/18/eu-turkey-statement/ (accessed September 3, 2018).

European Parliament. 2017. *All EU Countries Must Take Their Fair Share of Asylum Seekers*. October 19, 2017. www.europarl.europa.eu/news/en/press-room/20171016IPR86161/all-eu-countries-must-take-their-fair-share-of-asylum-seekers (accessed February 20, 2019).

European Union Agency for Fundamental Rights (FRA). 2010. *Police Stops and Minorities: Understanding and Preventing Discriminatory Ethnic Profiling*. Austria: FRA.

European Union Agency for Fundamental Rights (FRA). 2018. *FRA Work in the 'Hotspots'*. https://fra.europa.eu/en/theme/asylum-migration-borders/fra-work-hotspots (accessed February 3, 2019).

Fagioli-Ndlovu, Monica. 2015/12. *Somalis in Europe: INTERACT RP*. Robert Schuman Centre for Advanced Studies. Italy: San Domenico di Fiesole (FI): European University Institute.

Familiesammenføringskontoret. 2009. *Notat om praksis efter bestemmelsen i udlændingelovens § 9, stk. 8*. Udlændingestyrelsen. www.nyidanmark.dk/NR/rdonlyres/C6DCB443-2EAF-4CFA-8478-BE2293D71B87/0/notat_af_16102009_om_praksis_efter_bestemmelsen_i_udlaendingelovens_9_stk_8.pdf (accessed July 16, 2018).

Fassin, Didier. 2010. Inequality of lives, hierarchies of humanity: Moral commitments and ethical dilemmas of humanitarianism, pp. 238–255 in Ilana Feldman and Miriam Ticktin, eds. *In the Name of Humanity: The Government of Threat and Care*. Durham: Duke University Press.

Fassin, Didier. 2013. *Enforcing Order: An Ethnography of Urban Policing*. Cambridge: Polity Press.

Feldman, Gregory. 2011a. If ethnography is more than participant-observation, then relations are more than connections: The case for nonlocal ethnography in a world of apparatuses. *Anthropological Theory* 11(4): 375–395.

Feldman, Gregory. 2011b. *The Migration Apparatus: Security, Labor, and Policymaking in the European Union*. Stanford: Stanford University Press.

Feldman, Gregory. 2013. The specific intellectual's pivotal position: Action, compassion and thinking in administrative society, an Arendtian View. *Social Anthropology* 21(2): 135–154.

Feldman, Gregory. 2019. *The Gray Zone: Sovereignty, Human Smuggling, and Undercover Police Investigation in Europe*. Stanford: Stanford University Press.

Finn, Jonathan. 2005. Photographing fingerprints: Data collection and state surveillance. *Surveillance and Society* 3(1): 21–44.

Finn, Jonathan. 2009. *Capturing the Criminal Image: From Mug Shot to Surveillance Society*. Minneapolis: University of Minnesota Press.

Folketinget. 1996–97a. *§ 20-spørgsmål: Om DNA-undersøgelse af familiesammenførte somaliere*. Spm. Nr. S 1170. http://webarkiv.ft.dk/?/samling/arkiv.htm (accessed April 1, 2019).

Folketinget. 1996–97b. Retsudvalget, L 190 – bilag 29. *Betænkning afgivet af Retsudvalget den 26. maj 1997*. http://webarkiv.ft.dk/?/samling/arkiv.htm (accessed April 1, 2019).

Folketinget. 1997–98. Retsudvalget, L 59 – bilag 36. *Besvarelse af spørgsmål nr. 66 stillet af Folketingets Retsudvalg til indenrigsministeren den 5. maj 1998* (L 59 – bilag 22). http://webarkiv.ft.dk/?/samling/arkiv.htm (accessed April 1, 2019).

Folketingets Ombudsmand. 2008. *Udlændingemyndighedernes vejledning om familiesammenføring efter EU-retten mv. FOB nr. 08.238* www.ombudsmanden.dk/find/udtalelser/beretningssager/alle_bsager/2008-9-4/ (accessed March 24, 2019).

Folketingets Ombudsmand. 2019. *About the Ombudsman*. https://en.ombudsmanden.dk/introduktion/ (accessed March 24, 2019).

France 24. 2019, *March 1. Siege Ends After Deadly Al Shabaab Attack in Central Mogadishu*. www.france24.com/en/20190301-al-shabaab-terrorism-attack-mogadishu-stages-al-qaeda-bomb (accessed April 1, 2019).

Fratzke, Susan. 2015. *Not Adding Up: The Fading Promise of Europe's Dublin System. EU Asylum: Towards 2020 Project*. Washington, DC: Migration Policy Institute.

Frontex. 2006. *Amending Budget 2006*. https://frontex.europa.eu/assets/Key_Documents/Budget/Budget_2006.pdf (accessed April 5, 2019).

Frontex. 2007. *Amending Budget 2007 N2*. https://frontex.europa.eu/assets/Key_Documents/Budget/Budget_2007.pdf (accessed April 5, 2019).

Frontex. 2013. *Frontex Evaluation Report 2013 – Joint Operation METEOR 2013 – Air Border Cooperation*. Warsaw: Frontex.

Frontex. 2018. *Budget 2018*. https://frontex.europa.eu/assets/Key_Documents/Budget/Budget_2018.pdf (accessed April 5, 2019).

Frontex 2019a. *Vision, Mission and Values*. https://frontex.europa.eu/about-frontex/vision-mission-values/ (accessed August 2019)

Frontex. 2019b. *Origin & Tasks*. https://frontex.europa.eu/about-frontex/origin-tasks/ (accessed April 4, 2019).

Gammeltoft-Hansen, Thomas and Ninna Nyberg Sørensen. 2013. *The Migration Industry and the Commercialization of International Migration*. London: Routledge.

Gates, Kelly. 2011. *Our Biometric Future: Facial Recognition Technology and the Culture of Surveillance*. New York: New York University Press.

Gillespie, Marie, Souad Osseiran and Margie Cheesman. 2018. Syrian refugees and the digital passage to Europe: Smartphone infrastructures and affordances. *Social Media + Society*. January–March: 1–12.

Gold, Peter. 2000. *Europe or Africa?: A Contemporary Study of the Spanish North African Enclaves of Ceuta and Melilla*. Liverpool: Liverpool University Press.

Gorm-Hansen, Birgitte. 2011. Beyond the boundary: Science, industry, and managing symbiosis. *Bulletin of Science, Technology and Society* 31(6): 493–505.

Goudelis, Georgios, Anastasios Tefas and Ioannis Pitas. 2009. Emerging biometric modalities: A survey. *Journal on Multimodal User Interfaces* 2: 217–235.

Grasseni, Cristina. 2007a. Communities of Practice and Forms of Life: Towards a Rehabilitation of Vision? pp. 203–221 in Mark Harris, ed. *Ways of Knowing: Anthropological Approaches to Crafting Experience and Knowledge*. Oxford: Berghahn Books.

Grasseni, Cristina. 2007b. Good looking: Learning to be a cattle breeder, pp. 47–66 in Cristina Grasseni, ed. *Skilled Visions: Between Apprenticeship and Standards*. Oxford: Berghahn Books.

Grasseni, Cristina. 2010. Introduction, pp. 1–19 in Cristina Grasseni, ed. *Skilled Visions: Between Apprenticeship and Standards*. Oxford: Berghahn Books.

Grasseni, Cristina. 2011. Skilled visions: Toward an ecology of visual inscriptions, pp. 19–44 in Marcus Banks and Jay Ruby, eds. *Made to Be Seen: Perspectives on the History of Visual Anthropology*. Chicago: The University of Chicago Press.

Grasseni, Cristina. 2018. Skilled vision, pp. 1–7 in *The International Encyclopedia of Anthropology*. Oxford: John Wiley & Sons Ltd.

Grillo, Ralph. 2003. Cultural essentialism and cultural anxiety. *Anthropological Theory* 3(2): 157–173.

Grünenberg, Kristina. Forthcoming. *Wearing Someone Else's Face: Biometric technologies, anti-spoofing and the fear of the unknown*. Ms.

Gupta, Akhil and James Ferguson. 1997. Discipline and practice: 'The Field' as site, method, and location in anthropology, pp. 1–46 in Akhil Gupta and James Ferguson, eds. *Anthropological Locations: Boundaries and Grounds of a Field Science*. Berkeley: University of California Press.

Hansen, Peter. 2006. *Revolving Returnees: Meanings and Practices of Transnational Return Among Somalilanders*. PhD Thesis, Department of Anthropology, Faculty of Social Sciences, University of Copenhagen.

Haraway, Donna J. 2016. A Cyborg Manifesto: Science, technology, and socialist-feminism in the late twentieth century, pp. 3–90 in Cary Wolfe, ed. *Manifestly Haraway*. Minneapolis: University of Minnesota Press.

Harney, Nicholas. 2006. Rumour, migrants and the informal economies of Naples, Italy. *International Journal of Sociology and Social Policy* 26(9/10): 374–384.

Hartmann, Mia, Rosa Koss, Nadja Kirchhoff Hestehave, Lotte Høgh and Kira Vrist Rønn. 2018. Knowing from within. *Nordisk Politiforskning* 5(1): 7–27.

Hartung, Daniel, Martin A. Olsen, Xu Haiyun and Cristoph Busch. 2011. Spectral minutiae for vein pattern recognition. International Joint Conference on Biometrics (IJCB), pp. 1–7, Washington, DC, © 2011 IEEE. doi: 10.1109/IJCB.2011.6117549

Harvey, Penny and Soumhya Venkatesan. 2010. Faith, reason and the ethics of craftsmanship; Creating contingently stable worlds, pp. 129–143 in Matei Candea, ed. *The Social After Gabriel Tarde: Debates and Assessments*. London: Routledge.

Hastrup, Kirsten. 2004. Getting it right: Knowledge and evidence in anthropology. *Anthropological Theory* 4(4): 455–472.

Hastrup, Kirsten. 2015. *Speech at UCPH Anthropology's 70 Year Anniversary*. https://uniavisen.dk/en/speech-kirsten-hastrup-at-ucph-anthropologys-70-year-anniversary/ (accessed July 25, 2017).

Heinemann, Torsten, Ilpo Helén, Thomas Lemke, Ursula Naue and Martin G. Weiss. 2015. Constellations, complexities and challenges of researching DNA analysis for family reunification: An introduction, pp. 1–12 in Torsten Heinemann, Ilpo Helén, Thomas Lembke, Ursula Naue and Martin G. Weiss, eds. *Suspect Families: DNA Analysis, Family Reunification and Immigration Policies*. Farnham, Surrey: Ashgate.

Heinemann, Torsten and Thomas Lemke. 2014. Biological citizenship reconsidered: The use of DNA analysis by immigration authorities in germany. *Science, Technology & Human Values* 39(4): 488–510.

Heinemann, Torsten, Ursula Naue and Anna-Maria Tapaninen. 2013. Verifying the family? A comparison of DNA Analysis for family reunification in three European countries (Austria, Finland and Germany). *European Journal of Migration and Law* 15: 183–202.

Hertz, Michael. 2017. *Udlændingestyrelsen. Den Store Danske*. http://denstoredanske.dk/index.php?sideId=176269 (accessed January 20, 2019).

Hervik, Peter. 2011. *The Annoying Difference: The Emergence of Danish Neonationalism, Neoracism, and Populism in the Post-1989 World*. Oxford: Berghahn Books.

Herzfeld, Michael. 1993. *The Social Production of Indifference: Exploring the Symbolic Roots of Western Bureaucracy*. Chicago: The University of Chicago Press.

Hetherington, Kevin and Nick Lee. 2000. Social order and the blank figure. *Environment and Planning D: Society and Space* 18(2): 169–184.

Hoag, Colin. 2011. Assembling partial perspectives: Thoughts on the anthropology of bureaucracy. *PoLAR: Political and Legal Anthropology Review* 34(1): 81–94.

Hoffman, Steve G. 2017. Managing ambiguities at the edge of knowledge: Research strategy and artificial intelligence labs in an era of academic capitalism. *Science, Technology, & Human Values* 42(4): 703–740.

Hogle, Linda F. 2009. Pragmatic objectivity and the standardization of engineered tissues. *Social Studies of Science* 39(5): 717–742.

Holmberg, Lars. 2003. *Policing Stereotypes: A Qualitative Study of Police Work in Denmark.* Glienicke, Berlin: Galda + Wilch Verlag.

Holman Jones, Stacey and Ann Harris 2016. Traveling skin: A cartography of the body. *Liminalities: A Journal of Performance Studies* 12(1): 1–27.

Horst, Cindy. 2002. Vital links in social security: Somali refugees in the Dadaab Camps, Kenya. *Refugee Survey Quarterly* 21(1&2): 242–259.

Human Rights Watch. 2018. Somalia. Events of 2017. *World Report 2018.* www.hrw.org/world-report/2018/country-chapters/somalia (accessed February 2, 2019).

Hurwitz, Agnès. 1999. The 1990 Dublin convention: A comprehensive assessment. *International Journal of Refugee Law* 11(4): 646–677.

Hvidtfeldt, Camilla and Marie Louise Schultz-Nielsen. 2017. Flygtning og asylansøgere i Danmark 1992–2016. *Working Paper 50.* Copenhagen: Rockwoolfonden.

Ingold, Tim. 2000. *The Perception of the Environment: Essays on Livelihood, Dwelling and Skill.* London: Routledge.

Intagliata, Christopher. 2017. Biometrics identifies you in a heartbeat. *Scientific American.* October 6, 2017. www.scientificamerican.com/podcast/episode/biometric-identifies-you-in-a-heartbeat/ (accessed April 5, 2019).

Iriye, Akira. 1999. A century of NGOs. *Diplomatic History* 23(3): 421–435.

Jackson, Michael. 2002. Familiar and foreign bodies: A phenomenological exploration of the human-technological interface. *Journal of the Royal Anthropological Institute* 8(2): 333–346.

Jacobsen, Katja Lindskov. 2015. Experimentation in humanitarian locations: UNHCR and biometric registration of Afghan refugees. *Security Dialogue* 46(2): 144–164.

Jacobsen, Katja Lindskov. 2017. On humanitarian refugee biometrics and new forms of intervention. *Journal of Intervention and Statebuilding* 11(4): 529–551.

Jagd, Christina Bækkelund. 2007. *Medborger eller Modborger? Dansksomalieres kamp for at opbygge en meningsfuld tilværelse i det danske samfund – gennem et arbejde.* Ph.D. Thesis, University of Copenhagen. www.lev-dine-vaerdier.dk/CustomerData/Files/Folders/4-pdf/13_4.pdf (accessed April 1, 2019).

Jansen, Bram J. 2008. Between vulnerability and assertiveness: Negotiating resettlement in Kakuma Refugee Camp, Kenya. *African Affairs* 107(429): 569–587.

Jansen, Bram J. 2016. 'Digging Aid': The camp as an option in East and the Horn of Africa. *Journal of Refugee Studies* 29(2): 149–165.

Jaton, Florian. 2017. We get the algorithms of our ground truths: Designing: Referential databases in digital image processing. *Social Studies of Science* 47(6): 811–840.

Jöhncke, Steffen. 2011. Integrating Denmark: The welfare state as a national(ist) accomplishment, pp. 30–53 in Karen Fog Olwig and Karsten Paerregaard, eds. *The Question of Integration: Immigration, Exclusion and the Danish Welfare State.* Newcastle: Cambridge Scholars Publishing.

Jöhncke, Steffen and Vibeke Steffen. 2015. Introduction: Ethnographies on the limits of reason, pp. 9–39 in Vibeke Steffen, Steffen Jöhncke and Kirsten Marie Raahauge, eds. *Between Magic and Rationality: On the Limits of Reason in the Modern World.* Copenhagen: Museum Tusculanum Press.

Johnsdotter, Sara. 2015. European Somali Children dumped? On families, parents, and children in a transnational context. *European Journal of Social Work* 18(1): 81–96.

Kang, Mi-Jung. 2014. On digital photo-index, pp. 435–443 in Torkild Thellefsen and Bent Sorensen, eds. *Charles Sanders Peirce in His Own Words: 100 Years of Semiotics, Communication and Cognition.* Berlin and Boston: De Gruyter.

Kathrani, Paresh. 2010. Constructing human freedom: The refugee convention and networks of power. *Social Sciences Studies* 3(7): 115–124.

Kelly, Tobias. 2006. Documented lives: Fear and the uncertainties of law during second Palestinian intifada. *Journal of the Royal Anthropological Institute* 12(1): 89–107.

Kim, Jaeeun. 2011. Establishing identity: Documents, performance, and biometric information in immigration proceedings. *Law and Social Inquiry* 36(3): 760–786.

Kitchin, Robert and Martin Dodge. 2011. Rethinking maps, pp 108–114 in Robert Kitchin, Martin Dodge and Chris Perkins, eds. *The Map Reader: Theories of Mapping Practice and Cartographic Representation*. Hoboken: John Wiley & Sons.

Kleist, Nauja. 2004. Nomads, sailors and refugees: A century of Somali migration. *Sussex Migration Working Paper* 23: 1–14.

Kleist, Nauja. 2007. *Spaces of Recognition: An Analysis of Somali-Danish Associational Engagement and Diasporic Mobilization*. Ph.D. Thesis, University of Copenhagen.

Kleist, Nauja. 2016. Introduction: Studying hope and uncertainty in African Migration, pp. 1–20 in Nauja Kleist and Dorte Thorsen, eds. *Hope and Uncertainty in Contemporary African Migration*. London: Routledge.

Knorr-Cetina, Karin. 1981. *The Manufacture of Knowledge: An Essay on the Constructivist and Contextual Nature of Science*. Oxford: Pergamon Press.

Knorr-Cetina, Karin. 1995. Laboratory studies: The cultural approach to the study of science, pp. 140–166 in Sheila Jasanoff, Gerald E. Markle, James C. Peterson and Trevor Pinch, eds. *Handbook of Science and Technology Studies*. Thousand Oaks: Sage Publications.

Koch-Nielsen, Inger. 1996. Family obligations in Denmark. *The Danish National Institute of Social Research* 96: 3, Copenhagen. https://pure.sfi.dk/ws/files/332520/96_03.pdf (accessed March 15, 2018).

Kurihara, Yosuke and Kajiro Watanabe. 2011. Development of sensing device to detect persons hiding in a car. *IEEE Sensors Journal* 11(9): 1872–1878.

Kuster, Brigitta and Vassilis S. Tsianos. 2016. How to liquefy a body on the move: Eurodac and the making of the European digital border, pp. 45–63 in Raphael Bossong and Helena Carrapico, eds. *EU Borders and Shifting Internal Security: Technology, Externalization and Accountability*. Heidelberg: Springer.

Lammes, Sybille. 2017. Digital mapping interfaces: From immutable mobiles to mutable images. *New Media and Society* 19(7): 1019–1033.

Landsted, Anne Lea. 2018. Tamilsagen genfortalt: Dengang en bunke syltede sager væltede en hel regering. *Altinget*. January 10, 2018. www.altinget.dk/artikel/dengang-en-bunke-syltede-sager-vaeltede-en-hel-regering (accessed March 11, 2019).

Larkin, Brian. 2013. The politics and poetics of infrastructure. *Annual Review of Anthropology* 42: 327–343.

Latour, Bruno. 1983. Give me a laboratory, and I will raise the world, pp. 141–170 in Karin Knorr Cetina and Michael Mulkay, eds. *Science Observed: Perspectives on the Social Study of Science*. London and Beverly Hills: Sage Publications.

Latour, Bruno. 1987. *Science in Action*. Cambridge, MA: Harvard University Press.

Latour, Bruno. 1993. *We Have Never Been Modern*. Cambridge, MA: Harvard University Press.

Latour, Bruno. 1999. *Pandora's Hope: Essays on the Reality of Science Studies*. Cambridge, MA: Harvard University Press.

Latour, Bruno. 2010. *On the Modern Cult of the Factish Gods*. Durham: Duke University Press.

Latour, Bruno and Stephen Woolgar. 1986[1979]. *Laboratory Life: The Construction of Scientific Facts*. Princeton: Princeton University Press.

Law, John. 2011. Heterogeneous Engineering and Tinkering. *Heterogeneities.net*. September 19, 2011. www.heterogeneities.net/publications/Law2011HeterogeneousEngineeringAndTinkering.pdf (accessed April 14, 2017).

Levitt, Peggy, Jocelyn Viterna, Armin Mueller and Charlotte Lloyd. 2016. Transnational social protection: Setting the agenda. *Oxford Development Studies* 45(1): 2–19.

Lewis, Ioan Myrddin. 1993. *Understanding Somalia and Somaliland: Culture, History, Society*. London: Hurst and Company.

Lewis, Ioan Myrddin. 1994. *Blood and Bone: The Call of Kinship in Somali Society*. Lawrenceville: The Red Sea Press.

Lewis, Ioan Myrddin. [1965]2002. *A Modern History of the Somali*. Athens: Ohio University Press.

Leyden, John. 2002. Gummi bears defeat fingerprint sensors; Sticky problem for biometrics firms. *The Register; Biting the Hand that Feeds IT*. Situation Publishing. www.theregister.co.uk/2002/05/16/gummi_bears_defeat_fingerprint_sensors/ (accessed April 5, 2019).

Li, Stan Z. and Anil K. Jain. 2011. Introduction, pp. 1–15 in S. Z. Li and A. Jain, eds. *Handbook of Face Recognition*. London: Springer.

Lindley, Anna. 2010. *The Early Morning Phone Call: Somali Refugees' Remittances*. Oxford: Berghahn Books.

Liversage, Anika and Mikkel Rytter. 2015. Transnational marriages between closely related spouses in Denmark, pp. 130–152 in Alison Shaw and Aviad Raz, eds. *Cousin Marriages: Between Tradition, Genetic Risk and Cultural Change*. Oxford: Berghahn Books.

Lucht, Hans. 2012. *Darkness Before Daybreak: African Migrants Living on the Margins in Southern Italy Today*. Berkeley: University of California Press.

Luker, Trish. 2015. Performance anxieties: Interpellation of the refugee subject in law. *Canadian Journal of Law and Society* 30(1): 91–107.

MacDougall, David. 1998. Visual anthropology and the ways of knowing, pp. 61–92 in David MacDougall, ed. *Transcultural Cinema*. Princeton: Princeton University Press.

Maddox, Brenda. 2003. The double helix and the 'wronged heroine'. *Nature* 421: 407–408.

Madianou, Mirca and Daniel Miller. 2012. *Migration and the New Media: Transnational Families and Polymedia*. Abingdon: Routledge.

Magnet, Shoshana Amielle. 2011. *When Biometrics Fail: Gender, Race, and the Technology of Identity*. Durham: Duke University Press.

Maguire, Mark. 2009. The birth of biometric security. *Anthropology Today* 25(2): 9–14.

Maguire, Mark. 2014. Counter-terrorism in European airports, pp. 118–138 in Mark Maguire, Catarina Frois and Nils Zurawski, eds. *The Anthropology of Security: Perspectives From the Frontline of Policing, Counter-Terrorism and Border Control*. London: Pluto Press.

Maguire, Mark. 2018. Policing future crimes, pp. 137–158 in Mark Maguire, Ursula Rao and Nils Zurawski, eds. *Bodies as Evidence: Security, Knowledge, and Power*. Durham: Duke University Press.

Maguire, Mark, Catarina Frois and Nils Zurawski, eds. 2014. *Anthropology of Security: Perspectives From the Frontline of Policing, Counter-Terrorism and Border Control*. Chicago: The University of Chicago Press.

Maltoni, Davide, Dario Maio, Anil K. Jain and Salil Prakhakar. 2009. Introduction, pp. 1–56 in Davide Maltoni, Dario Maio, Anil K. Jain and Salil Prakhakar, eds. *Handbook of Fingerprint Recognition*. London: Springer.

Matsumoto, Tsutomo, Hiroyuki Matsumoto, Koji Yamada and Satoshi Hoshino. 2002. Impact of artificial gummy fingers on fingerprint systems. *Proceedings of SPIE 4677, Optical Security and Counterfeit Deterrence Techniques IV (19 April 2002).*

Mehlsen, Camilla, Anne Vindum og Ane Nordentoft. 2015. Familien under forandring. *Faktalink.* https://faktalink.dk/titelliste/familien-under-forandring (accessed October 28, 2018).

MetroXpress. 2016. *Fanget i Jordan på grund af DNA-test: Savner du os ikke, far?* www.mx.dk/feature/cover/story/22752963 (accessed August 31, 2016).

Mitchell, William J. T. 2002. Showing seeing: A critique of visual culture. *Journal of Visual Culture* 1(2): 165–181.

Mogensen, Hanne. 2011. Caught in the grid of difference and gratitude: HIV positive Africans facing the challenges of Danish sociality, pp. 207–229 in Karen Fog Olwig and Karsten Paerregaard, eds. *The Question of Integration: Immigration, Exclusion and the Danish Welfare State.* UK: Cambridge Scholars.

Møhl, Perle. 1993. Forførelsens element – når vi ser etnografiske film. *Tidsskriftet Antropologi* 27: 17–27.

Møhl, Perle. 1997. *Village Voices: Coexistence and Communication in a Rural Community in Central France.* Copenhagen: Museum Tusculanum Press.

Møhl, Perle. 2011. Mise-En-Scène, knowledge and participation: Considerations of a filming anthropologist. *Visual Anthropology* 24(3): 227–245.

Møhl, Perle. 2012. Cartographic tours, pp. 73–116 in Perle Møhl, ed. *Omens and Effect: Divergent Perspectives on Emerillon Time, Space and Existence.* Meaulne: Semeïon Editions.

Møhl, Perle. 2018. Border control and blurred responsibilities at the Airport, pp. 118–135 in Tessa G. Diphoorn and Erella Grassiani, eds. *Security Blurs, the Politics of Plural Security Provision.* London: Routledge.

Møhl, Perle. Forthcoming. *Biometric Technologies, Data and the Sensory Work of Border Control.* Ms.

Møhl, Perle and Nanna Hauge Kristensen. 2018. At bruge billeder og lyd: Sensorisk antropologi, pp. 227–244 in Helle Bundgaard, Hanne Mogensen and Cecilie Rubow, eds. *Antropologiske Projekter: En Grundbog.* København: Samfundslitteratur.

Mosco, Vincent. 2004. *The Digital Sublime: Myth, Power, and Cyberspace.* Cambridge, MA: MIT Press.

Mosse, David. 2008. Anti-social anthropology? Objectivity, objection, and the ethnography of public policy and professional communities. *Journal of the Royal Anthropological Institute (N.S.)* 12: 935–956.

Muižnieks, Nils. 2017. Ending restrictions on family reunification: Good for refugees, good for host societies. *Council of Europe, Commissioner for Human Rights, The Commissioner's Human Rights Comments.* www.coe.int/en/web/commissioner/-/ending-restrictions-on-family-reunification-good-for-refugees-good-for-host-societies?desktop=true (accessed October 3, 2018).

Mutiga, Murithi and Emma Graham-Harrison. 2016. Kenya says it will shut down world's biggest refugee camp Dadaab. *The Guardian.* May 11, 2016. www.theguardian.com/world/2016/may/11/kenya-close-worlds-biggest-refugee-camp-dadaab (accessed March 3, 2019).

Nail, Thomas. 2015. *The Figure of the Migrant.* Stanford: Stanford University Press.

Navaro-Yashin, Yael. 2007. Make-believe papers, legal forms and the counterfeit: Affective interactions between documents and people in Britain and Cyprus. *Anthropological Theory* 7(1): 79–98.

Netz, Sabine. Forthcoming. *Teeth and Truth? Age IDentities of Migrants in the Making.* Ms.

Nielsen, Katrine Bang. 2004. Next Stop Britain: The influence of transnational networks on the secondary movement of Danish Somalis. *Sussex Migration Working Paper No. 22*, Sussex Centre for Migration Research. www.sussex.ac.uk/webteam/gateway/file.php?name=mwp22.pdf&site=252 (accessed April 1, 2019).

Norman, Karin. 2005. The workings of uncertainty: Interrogating cases on refugees in Sweden. *Social Analysis* 49(3): 189–214.

Nye, David. 1994. *American Technological Sublime*. Cambridge, MA: MIT Press.

Okely, Judith. 2001. Visualism and landscape: Looking and seeing in Normandy. *Ethnos* 66(1): 99–120.

Olwig, Karen Fog. 1997. Cultural sites: Sustaining a home in a deterritorialized world, pp. 17–38 in Karen Fog Olwig and Kirsten Hastrup, eds. *Siting Culture: The Shifting Anthropological Object*. London: Routledge.

Olwig, Karen Fog. 2007. *Caribbean Journeys: An Ethnography of Migration and Home in Three Family Networks*. Durham: Duke University Press.

Olwig, Karen Fog. 2015. The duplicity of diversity: Caribbean immigrants in Denmark. *Ethnic and Racial Studies* 38(7): 1104–1119.

Olwig, Karen Fog. 2018. Migration as adventure: Narrative self-representation among Caribbean migrants in Denmark. *Ethnos* 83(1): 156–171.

Olwig, Karen Fog. In Press. The right to a family life and the biometric 'truth' of family reunification: Somali refugees in Denmark. *Ethnos*.

Olwig, Karen Fog. Forthcoming. The end of flight? Temporariness, uncertainty and meaning in refugee life. Ms.

Olwig, Karen Fog, Birgitte Romme Larsen and Mikkel Rytter, eds. 2012. *Migration, Family and the Welfare State: Integrating Immigrants and Refugees in Scandinavia*. London: Routledge.

Olwig, Karen Fog and Karsten Paerregaard, eds. 2011. *The Question of Integration: Immigration, Exclusion and the Danish Welfare State*. Newcastle: Cambridge Scholars Publishing.

Olwig, Karen Fog and Ninna Nyberg Sørensen. 2002. Mobile livelihoods: Making a living in the world, pp. 1–19 in N. N. Sørensen and K. F. Olwig, eds. *Work and Migration: Life and Livelihoods in a Globalizing World*. London: Routledge.

Olwig, Kenneth R. 2002. *Landscape, Nature and the Body Politic: From Britain's Renaissance to America's New World*. Madison: University of Wisconsin Press.

Olwig, Kenneth R. 2004. This is not a landscape: Circulating reference and land shaping, pp. 41–66 in Hannes Palang, Helen Sooväli, Marc Antrop and Gunhild Setten, eds. *European Rural Landscapes: Persistence and Change in a Globalising Environment*. Dordrecht: Kluwer.

Olwig, Kenneth R. 2013. Heidegger, Latour and the reification of things: The inversion and spatial enclosure of the substantive landscape of things – The Lake District case. *Geografiska Annaler Series B: Human Geography* 95(3): 251–273.

Olwig, Kenneth R. 2018. England's 'Lake District' and the 'North Atlantic Archipelago': A body of managed land contra a body politic. *Landscape Research* 43(8): 1032–1044.

Orsini, Giacomo, Andrew Canessa, Luis Gonzaga Martínez del Campo and Jennifer Ballantine Pereira. 2017. Fixed Lines, permanent transitions. International Borders, cross-border communities and the transforming experience of otherness. *Journal of Borderlands Studies*. July: 1–16.

O'Sullivan, Kevin, Matthew Hilton and Juliano Fiori. 2016. Humanitarianisms in context. *European Review of History* 23(1–2): 1–15.

Pack, Sasha D. 2014. The making of the Gibraltar-Spain border: Cholera, contraband, and spatial reordering, 1850–1873. *Mediterranean Historical Review* 29(1): 71–88.

Pallister-Wilkins, Polly. 2017. The tensions of the Ceuta and Melilla border fences, pp. 63–81 in Paolo Gaibazzi, Stephan Dünnwald and Alice Bellagamba, eds. *EurAfrican Borders and Migration Management*. New York: Palgrave Macmillan US.

Palm, Anja. 2017. *The Italy-Libya Memorandum of Understanding: The Baseline of a Policy Approach Aimed at Closing All Doors to Europe?* http://eumigrationlawblog.eu/the-italy-libya-memorandum-of-understanding-the-baseline-of-a-policy-approach-aimed-at-closing-all-doors-to-europe/ (accessed October 3, 2018).

Palsson, Gisli. 1994. Enskilment at sea. *Man (N.S.)* 29(4): 901–927.

Pedersen, Morten Axel. 2011. *Not Quite Shamans: Spirit Worlds and Political Lives in Northern Mongolia*. Ithaca: Cornell University Press.

Peirce, Charles S. 1998. *The Essential Peirce: Selected Philosophical Writings, Vol. 2 (1893–1913)*. Bloomington: University of Indiana Press.

Pickering, Sharon and Leanne Weber. 2006. Borders, mobility and technologies of control, pp. 1–19 in S. Pickering and L. Weber, eds. *Borders, Mobility and Technologies of Control*. Dordrecht: Springer.

Pinney, Christopher. 2008. The prosthetic eye: Photography as cure and poison. *Journal of the Royal Anthropological Institute* 14(1): 33–46.

Politiken. 2011. *Udskældt Udlændingeservice skal have et nyt navn*. December 23, 2011. https://politiken.dk/indland/art5038316/Udsk%C3%A6ldt-Udl%C3%A6ndingeservice-skal-have-nyt-navn (accessed January 20, 2019).

Rabinow, Paul. 2004. Anthropologie des Zeitgenössischen, pp. 56–65 in P. Rabinow, ed. *Anthropologie der Vernunft*. Frankfurt am Main: Suhrkamp.

Ramani, Madhvi. 2018, September 24. How France created the metric system. Places that changed the world. *BBC Travel Service*. www.bbc.com/travel/story/20180923-how-france-created-the-metric-system (accessed April 1, 2019).

Rapiscan. 2017. *Threat Image Projection*. www.rapiscansystems.com/en/products/radiation_detection/productsrapiscan_threat_image_projection (accessed December 4, 2017).

Rapport, Nigel. 2013. Opportunity, cliché and cosmopolitan politesse: A commentary on Feldman. *Social Anthropology* 21(2): 157–163.

Rapport, Nigel. 2017. Towards a cosmopolitan anthropology of anyone. *Sites* 14(2): 1–13.

Reeves, Madeleine. 2013. Clean fake: Authenticating documents and persons in migrant Moscow. *American Ethnologist* 40(3): 508–524.

Restrup, Anne Katrine. 2014. Der er 37 versioner af familien Danmark. *Kristeligt Dagblad*. September 5, 2014. www.kristeligt-dagblad.dk/liv-sj%C3%A6l/2014-09-05/s%C3%A5dan-ser-familien-danmark-ud (accessed October 28, 2018).

Rheinberger, Hans Jörg. 2009. Recent science and its exploration: The case of molecular biology. *Studies in History and Philosophy of Biological and Biomedical Sciences* 40(1): 6–12.

Richards, Anne R. 2003. Argument and authority in the visual representations of science. *Technical Communication Quarterly* 12(2): 183–206.

Richter, Line. 2018. *Gaps in a Bordered World: Malian Men Trying to Make It to and in Europe*. PhD Dissertation, University of Copenhagen, Denmark.

Riedel, Lisa and Gerald Schneider. 2017. Dezentraler Asylvollzug diskriminiert: Anerkennungsquoten von Flüchtlingen im bundesdeutschen Vergleich, 2010–2015. *PVS Politische Vierteljahresschrift*, 58(1): 21–48.

Rinaldi, Andrea. 2016. Biometrics' new identity – Measuring more physical and biological traits. *Science & Society, EMBO Reports* 17(1): 22–26.

Rodrigues, Laurie. 2012. Seeing immanent difference: Lorna Simpson and the face's affect. *Rhizomes* 23. www.rhizomes.net/issue23/rodrigues/rodrigues.html (accessed August 15, 2018).

Rytter, Mikkel. 2011. 'The family of Denmark' and 'the Aliens': Kinship images in Danish integration politics, pp. 54–76 in Karen Fog Olwig and Karsten Paerregaard, eds. *The Question of Integration: Immigration, Exclusion and the Danish Welfare State.* Newcastle: Cambridge Scholars Publishing.

Rytter, Mikkel. 2013. *Family Upheaval: Generation, Mobility and Relatedness Among Pakistani Migrants in Denmark.* Oxford: Berghahn Books.

Rytter, Mikkel. 2018. Writing against integration: Danish imaginaries of culture, race and belonging. *Ethnos* 84(4): 678–697.

Rytter, Mikkel and Karen Fog Olwig. 2011. Introduction: Family, religion and migration in a global world, pp. 9–26 in Mikkel Rytter and Karen Fog Olwig, eds. *Mobile Bodies, Mobile Souls: Family, Religion and Migration in a Global World.* Aarhus: Aarhus University Press.

Saddiki, Said. 2012. Les Clôtures de Ceuta et de Melilla. *Études Internationales* 43(1): 49.

Saddiki, Said. 2017. The fences of Ceuta and Melilla, pp. 57–81 in *World of Walls: The Structure, Roles and Effectiveness of Separation Barriers.* Cambridge: Open Book Publishers.

Sahlins, Marshall. 2013. *What Kinship Is – And Is Not.* Chicago: University of Chicago Press.

Salter, Mark B. 2013. To make move and let stop: Mobility and the assemblage of circulation. *Mobilities* 8(1): 7–19.

Sanchez del Rio, Jose, Daniela Moctezuma, Cristina Conde, Isaac Martin de Diego and Enrique Cabello. 2016. Automated border control E-Gates and facial recognition systems. *Computers and Security* 62: 49–72.

Scheel, Stephan and Philipp Ratfisch. 2014. Refugee protection meets migration management: UNHCR as a global police of populations. *Journal of Ethnic and Migration Studies* 40(6): 924–941.

Scherer, Steve. 2015, 10 June. Migrants race through Italy to dodge EU asylum rules. *Reuters.* www.reuters.com/article/us-europe-migrants-asylum-insight/migrants-race-through-italy-to-dodge-eu-asylum-rules-idUSKBN0OQ0EU20150610 (accessed February 14, 2019).

Schindel, Estela. 2016. Bare life at the European Borders. Entanglements of technology, society and nature. *Journal of Borderlands Studies* 31(2): 219–234.

Schmeling, Andreas, Pedro Manuel Garamendi, Jose Luis Prieto and María Irene Landa. 2011. Forensic age estimation in unaccompanied minors and young living adults, forensic medicine – From old problems to new challenges, in Prof. Duarte Nuno Vieira, ed. *InTech.* www.intechopen.com/books/forensic-medicine-from-old-problems-to-new-challenges/forensic-ageestimation-in-unaccompanied-minors-and-young-living-adults (accessed April 1, 2019).

Schmidt, Garbi, 2011. Law and identity: Transnational arranged marriages and the boundaries of Danishness. *Journal of Ethnic and Migration Studies* 37(2): 257–275.

Schrooten, Mieke, Noel B. Salazar and Gustavo Dias. 2016. Living in mobility: Trajectories of Brazilians in Belgium and the UK. *Journal of Ethnic and Migration Studies* 42(7): 1199–1215.

Schuster, Liza. 2005. The continuing mobility of migrants in Italy: Shifting between places and statuses. *Journal of Ethnic and Migration Studies* 31(4): 757–774.

Schuster, Liza. 2011. Dublin 2 and Eurodac: Examining the (un)intended(?) consequences. *Gender, Place and Culture* 18(3): 401–416.

Schwaninger, Adrian and Franziska Hofer. 2005. Using threat image projection data for assessing individual screener performance. *WIT Transactions on the Built Environment* 82: 417–426.

Scott, James C. 1998. *Seeing Like a State: How Certain Schemes to Improve the Human Condition Have Failed*. New Haven: Yale University Press.

Shackleton, Mark, Kally Yuen, Andrew F. Little, Stephen Schlicht and Sue-Anne McLachlan. 2004. Reliability of X-rays and bone scans for the assessment of changes in skeletal metastases from breast cancer. *Internal Medicine Journal* 34(11): 615–620.

Shore, Cris and Susan Wright. 2015. Audit culture revisited. *Current Anthropology* 56(3): 421–444.

Silverstein, Jason. 2012. Bonds beyond blood. DNA Testing and refugee family reunification. *Anthropology News*. April: 11. www.anthropology-news.org (accessed December 15, 2018).

Simonsen, Anja. 2017. *Tahriib: The Journey Into the Unknown: An Ethnography of Mobility, Insecurities and Uncertainties Among Somalis en Route*. Ph.D. Dissertation, University of Copenhagen, Denmark.

Simonsen, Anja. 2018. Migrating for a better future: 'Lost time' and its social consequences among young Somali migrants, pp. 103–122 in Pauline Gardiner Barber and Winnie Lem, eds. *Migration, Temporality and Capitalism: Entangled Mobilities Across Global Spaces*. Switzerland: Palgrave Macmillan.

Skak, Tine. 1998. De er så svære at integrere, pp. 96–119 in Ann-Belinda S. Preis, ed. *Kan vi leve sammen? Integration mellem politik og praksis*. Copenhagen: Munksgaard.

Skytte, Marianne. 2007. *Etniske minoritetsfamilier og socialt arbejde*. København: Hans Reitzels Forlag.

Sølvsten, Charlotte. 2016. Politikere forargede: Børnene skal DNA-testes. *TVSYD*. May 26, 2016. www.tvsyd.dk/artikel/politikere-forargede-boernene-skal-dna-testes (accessed August 31, 2016).

Sommer, Matthew H. 2015. *Polyandry and Wife-Selling in Qing Dynasty China: Survival Strategies and Judicial Interventions*. Berkeley: University of California Press.

Sørensen, Ninna Nyberg and Thomas Gammeltoft-Hansen. 2013. Introduction, pp. 1–23 in Thomas Gammeltoft-Hansen and Ninna Nyberg Sørensen, eds. *The Migration Industry and the Commercialization of International Migration*. Abingdon: Routledge.

Strathern, Marilyn, ed. 2000. *Audit Cultures: Anthropological Studies in Accountability, Ethics and the Academy*. London and New York: Routledge.

Strickland, Eliza. 2011. Can biometrics ID and identical twin? Notre Dame researchers put face-recognition software to the 'torture test': Genetically identical people. *IEEE Spectrum*.

Suskind, Ron. 2006. *The One Percent Doctrine: Deep Inside America's Pursuit of Its Enemies Since 9/11*. New York: Simon and Schuster.

Svendsen, Mette N. 2011. Articulating potentiality: Notes on the delineation of the blank figure in human embryonic stem cell research. *Cultural Anthropology* 26(3): 414–437.

Tapaninen, Anna-Maria, Miia Halme-Tuomisaari and Viljami Kankaanpää. 2019 Mobile lives, immutable facts: Family reunification of children in Finland. *Journal of Ethnic and Migration Studies* 45(5): 825–841.

Thévenot, Laurent. 2009. Postscript to the special issue: Governing life by standards: A view from engagements. *Social Studies of Science* 39(5): 793–813.

Thomson, Marnie Jane. 2012. Black boxes of bureaucracy: Transparency and opacity in the resettlement process of Congolese refugees. *PoLAR. Political and Legal Anthropology Review* 35(2): 186–205.

Tome, Pedro and Sebastién Marcel. 2015. On the vulnerability of palm vein recognition to spoofing attacks. *International Conference on Biometrics (ICB)*: 319–325.

Topak, Özgun E. 2014. The biopolitical border in practice: Surveillance and death at the Greece-Turkey borderzones. *Environment and Planning D: Society and Space* 32(5): 815–833.

Torpey, John C. 2018. *The Invention of the Passport: Surveillance, Citizenship and the State*. Cambridge: Cambridge University Press.

Traweek, Sharon. 1988. *Beamtimes and Lifetimes: The World of High-Energy Physicists*. Cambridge, MA: Harvard University Press.

Turnbull, Colin. 2010. Maps and plans in 'Learning to See': The London Underground and Chartres Cathedral as examples of performing design, pp. 125–141 in Cristina Grasseni, ed. *Skilled Visions: Between Apprenticeship and Standards*. Oxford: Berghahn Books.

Udlændinge- og Integrationsministeriet. 2018. Udlændingestyrelsen. http://uim.dk/us (accessed January 18, 2018).

Udlændingenævnet. 2014. *2. Beretning – 2014*. Formandskabet. http://udln.dk/~/media/raw_udln/Aarsberetning_2014_1_udkast_2.ashx (accessed March 16, 2018).

Udlændingestyrelsen. 2016. Udlændingestyelsen undersøger opholdstilladelser til somaliere, https://nyidanmark.dk/da/News%20Front%20Page/2017/12/US_undersoeger_ophold-stilladelser_somaliere (accessed February 12, 2019).

UNHCR. 2019a. *Europe Situation*. www.unhcr.org/europe-emergency.html (accessed April 5, 2019).

UNHCR. 2019b. *Global Trends: Forced Displacement in 2016*. www.unhcr.org/globaltrends 2016/ (accessed April 5, 2019).

UNHCR, The United Nations Refugee Agency. 2017. *Global Trends: Forced Displacement Trends in 2017*. www.unhcr.org/5b27be547.pdf (accessed February 14, 2019).

United Nations. 1948. *Universal Declaration of Human Rights*. www.un.org/en/universal-declaration-human-rights/ (accessed December 20, 2018).

United Nations Refugee Agency. 2012. *Refugee Family Reunification: UNHCR's Response to the European Commission Green Paper on the Right to Family Reunification of Third Country Nationals Living in the European Union* (Directive 2003/86/EC). UN High Commissioner for Refugees (UNHCR). www.refworld.org/docid/4f55e1cf2.html (accessed March 14, 2019).

van der Ploeg, Irma. 1999. The illegal body: 'Eurodac' and the politics of biometric identification. *Ethics and Information Technology* 1(4): 295–302.

van der Ploeg, Irma. 2003. Biometrics and privacy: A note on the politics of theorizing technology. *Information, Communication and Society* 6(1): 85–104.

van Gennep, Arnold. 1960[1908]. *The Rites of Passage*. Chicago: The University of Chicago Press.

Vrbančić, Mario. 2005. Burroughs phantasmic maps. *New Literary History* 36(2): 313–326. Baltimore, MD: Johns Hopkins University Press.

Wahlberg, Ayo. 2018. *Good Quality: The Routinization of Sperm Banking in China*. Berkeley: University of California Press.

Watters, Charles. 2007. Refugees at Europe's Borders: The moral economy of care. *Transcultural Psychiatry* 44(3): 394–417.

Weitzberg, Keren. 2017. *We Do Not Have Borders: Greater Somalia and the Predicaments of Belonging in Kenya*. Athens: Ohio University Press.

Wells, Joshua J. 2014. Keep calm and remain Human: How we have always been Cyborgs and theories on the technological present of anthropology. *Reviews in Anthropology* 43(1): 5–34.

Wilson, Dean. 2006. Biometrics, borders and the ideal suspect, pp. 87–109 in Sharon Pickering and Leanne Weber, eds. *Borders, Mobility and Technologies of Control*. Dordrecht: Springer.

Winston, Brian and Hing Tsang. 2009. The subject and the indexicality of the photograph. *Semiotica* 1/4(173): 453–469.

Xiang, Biao and Johan Lindquist. 2014. Migration infrastructure. *International Migration Review* 48(1): 122–148.

XOVIS AG. 2019. The digital transformation of passenger flows and queues. *Future Airport* 28. www.futureairport.com/contractors/passenger-handling/xovis/ (accessed January 2, 2019).

Yapijakis, Christos. 2017. Ancestral concepts of human genetics and molecular medicine in epicurean philosophy, pp. 41–57 in Heike I. Petermann, Peter S. Harper and Susanne Doetz, eds. *History of Human Genetics*. Berlin: Springer.

Yuksel, Aycan, Lale Akarun and Bülent Sankur. 2011. Hand vein biometry based on geometry and appearance methods. *IET Computer Vision* 5(6): 398–406.

Zijlstra, Judith and Ilse van Liempt. 2017. Smart(phone) travelling: Understanding the use and impact of mobile technology on irregular migration journeys. *International Journal of Migration and Border Studies* 3(2/3): 174–191.

Index

For Product Safety Concerns and Information please contact our EU
representative GPSR@taylorandfrancis.com
Taylor & Francis Verlag GmbH, Kaufingerstraße 24, 80331 München, Germany

www.ingramcontent.com/pod-product-compliance
Lightning Source LLC
Chambersburg PA
CBHW060359220326
41598CB00023B/2971